ISBN 978-1-330-29223-5
PIBN 10015969

1 MONTH OF
FREE
READING

at
www.ForgottenBooks.com

By purchasing this book you are eligible for one month membership to ForgottenBooks.com, giving you unlimited access to our entire collection of over 1,000,000 titles via our web site and mobile apps.

To claim your free month visit:
www.forgottenbooks.com/free15969

English
Français
Deutsche
Italiano
Español
Português

www.forgottenbooks.com

Mythology Photography **Fiction**
Fishing Christianity **Art** Cooking
Essays Buddhism Freemasonry
Medicine **Biology** Music **Ancient**
Egypt Evolution Carpentry Physics
Dance Geology **Mathematics** Fitness
Shakespeare **Folklore** Yoga Marketing
Confidence Immortality Biographies
Poetry **Psychology** Witchcraft
Electronics Chemistry History **Law**
Accounting **Philosophy** Anthropology
Alchemy Drama Quantum Mechanics
Atheism Sexual Health **Ancient History**
Entrepreneurship Languages Sport
Paleontology Needlework Islam
Metaphysics Investment Archaeology
Parenting Statistics Criminology
Motivational

BOTANY ALL THE YEAR ROUND

A PRACTICAL TEXT-BOOK FOR SCHOOLS

BY

E. F. ANDREWS

HIGH SCHOOL, WASHINGTON, GEORGIA

———oo⁙oo———

NEW YORK ∴ CINCINNATI ∴ CHICAGO

AMERICAN BOOK COMPANY

1903.

PREFACE

Most of the recent text-books of botany, excellent as many of them are, fail to meet the conditions of the average public school, where expensive laboratory appliances are out of the question, and time to make a proper use of them is equally unattainable. It is one of the anomalies of our educational system that the study of plants, if provided for at all, should be confined mainly to city schools, where it is necessarily carried on under disadvantageous conditions, while it is almost entirely neglected in the country, where the great laboratory of nature stands invitingly open at every schoolhouse door.

The writer believes that this neglect is largely due to the want of a text-book suited to general use, in which the subject is treated in a manner at once simple, practical, and scientific. It is with a desire to meet this need and to encourage a more general adoption of botanical studies in the public schools that the present work has been undertaken. It aims, in the first place, to lead the pupil to nature for the objects of each lesson ; and in the second place, to provide that the proper material shall be always available by so arranging the lessons that each subject will be taken up at just the time of the year when the material for it is most abundant. In this way the study can be carried on all the year round, a plan which will be found much better than crowding the whole course into a few weeks of the spring term.

In order to provide for this all the year round course it has been necessary to depart somewhat from the usual order of arrangement, but years of experience have convinced the writer that the advantages to be gained by having fresh material always at hand are sufficient to outweigh other considerations that might be advanced in favor of established methods. The leaf has been selected as the starting point mainly because it is the most convenient material at hand in September, when the schools begin; and it is such an important and fundamental part of the plant that a thorough acquaintance with its nature and functions will clear the way to an understanding of many of the problems that will face the student later.

It is not expected that all the work outlined in the book will be done just as it is written, and much of it may even have to be omitted altogether. Each teacher can select such parts as are suited to the circumstances of his school, passing lightly· over some topics, giving more attention to others, as material and opportunity may suggest. The study of botany is necessarily sectional to some extent, because nature is so, but the method here outlined is of universal application and every teacher can select his own specimens in accordance with the directions given in the body of the book. Prominence is given to the more familiar forms of vegetation presented by the seed-bearing plants, as the author believes that for ordinary purposes the best results are to be obtained by proceeding from the familiar and well known to the more primitive and obscure forms. The reverse order may be better for the trained investigator; the other is simpler and more attractive, and for ordinary purposes the only practicable one. The average boy and girl will learn more of what it concerns them to know about stem structure, for instance, from a cornstalk, and a handful of chips, or even from the graining of the timber out of which their desks are made, than from the most elaborate study of the *xylem* and the *phloem* and the *collenchymatous tissues*. For we must bear in mind that the object of teaching botany in the common schools is not to train experts and investigators but intelligent observers.

In giving the botanical names of plants the terminology of Gray's handbooks is adhered to, partly because they are at present the most generally available for school use, and more especially because the new terminology is in such an unsettled state that nobody can say what it will be to-morrow or next day. Hence, while recognizing the desirability of some of the changes proposed, the author does not think it advisable to confuse the beginner by introducing him to a system that is undergoing a period of transition. After all, this is a mere matter of names, and does not affect the point that ought to be kept in view — the hereditary relationships of plants.

The experiments described are for the most part very simple, requiring no appliances but such as the ingenuity of the teacher and pupils can easily devise, as will be seen by a glance at the list on pages 12 and 13 of the text.

Teachers trained in normal schools, where all the material needed for their work is furnished by the State, and ample time allowed them, are often completely at a loss when transferred to country schools, where no provision is made for laboratory work, and the patrons grumble if called upon to buy so much as a drawing book or a hand lens. Too often they can think of no other resource than to drop botany from the curriculum altogether rather than depart from what they have been taught to consider the only scientific method. It is hoped that the present volume may suggest a better way out of the difficulty, and also that it may be a help to those who have not enjoyed the advantage of a technical training.

The writer would not underrate the value of histological studies or the advantages of a well equipped laboratory, but since these are at present clearly out of the reach of the great majority of the school population, and more especially of that very class to whom the study of plants is of the greatest practical importance and into whose lives it would bring the greatest amount of pleasure and of intellectual enlargement, it has been made the aim of this book to show that botany can be taught to some purpose by means within the reach of everybody. It has also been the author's aim to keep constantly in view the intimate relations between botany and agriculture. The practical questions at the end of each section, it is hoped, will have the effect of bringing out these relations more clearly and at the same time of leading the pupil to reason for himself and draw his own inferences from the common phenomena about him.

The author takes pleasure in acknowledging here the many obligations due to Dr. C. O. Townsend of the United States Department of Agriculture for his very effective assistance in revising the manuscript of this work; also to Professor Charles Wright Dodge of the University of Rochester, and Professor W. F. Ganong of Smith College for valuable criticisms and suggestions. Acknowledgments are also due to Messrs. D. Appleton and Company, for permission to use illustrations from Coulter's "Plant Relations" and "Plant Structures," copyright, 1899, and to the owners of Gray's Botanies, to Professor William Trelease of the Missouri Botanical Garden, to Mr. Gifford Pinchot of the United States Department of Agriculture, and to Mr. W. S. Bailey of the Chautauqua Bureau of Publication, for permission to reproduce illustrations from their publications. Quite a number of the figures used are from original drawings by pupils of the Washington, Ga., High School.

CONTENTS

BOTANY ALL THE YEAR ROUND

——oo;o;oo——

I. INTRODUCTION

1. General Statement. — Botany is the science which treats of the vegetable kingdom, but the subject is so comprehensive that it has been divided into many branches, each of which is a science in itself. For instance, there are Mycology, the study of mushrooms and other fungi; Bacteriology, the study of the microscopic forms concerned in the process of fermentation, and in the production of disease; Paleobotany, the study of fossil plants — and many others, with which we have no concern at present. Each of these studies may be viewed under various aspects, and these in turn have given rise to still other divisions of the subject, such as, —

2. Morphology, or Structural Botany, the study of the different organs or parts of plants in regard to their form and uses and the various changes and adaptations they may undergo.

3. Histology, or Plant Anatomy, the microscopic study of the minute structure of plant organs. This can not be carried on well without the use of the compound microscope and other appliances not obtainable in many schools. Something, however, may be learned from a few simple experiments, accompanied by intelligent observation with a hand lens, and it is only in so far as it can be carried on by ordinary means like these, that this branch of the subject is touched upon in the present work.

4. Vegetable Physiology, the study of the action of living plants and their organs, their mode of growth and reproduction, and their various movements for adjustment to their surroundings, as the attraction of roots toward moisture and of leaves toward light.

5. Ecology, the study of plants in their relations to external conditions, or, to use a more convenient term, their environment. This is one of the most interesting and important of all the departments of botany, and presents many points of direct practical concern to the farmer.

6. Taxonomy, called also Systematic or Descriptive Botany, the study of plants in their relationships to one another. Its work is to note their resemblances and differences, and by means of these to classify or distribute them into certain great groups called families or orders, and these again into lesser groups of genera and species. This work of classification was formerly considered the chief end of the study of botany, which thus too often degenerated into a mere mechanical drill in hunting down plants and labeling them with hard names. The tendency at present is to ignore this part of the subject altogether, which is nearly as great a mistake as the old-fashioned error of thinking that the study of botany consisted merely in learning a string of hard words. One of the chief pleasures to be derived from botanical studies, for most of us, consists in being able to know and recognize the various plants we meet with. The first thing we all ask on seeing a new shrub or flower is, "What is it?" and this question can be answered satisfactorily only by referring each to its proper class or order.

7. Learn to know the Common Plants. — These five subdivisions make up the study of botany, as generally taught in the schools. They apply to all plants, and the only practicable way for most of us to learn them is by a study of the common vegetable life about us.

8. Definitions. — "Organ" is a general name for any part of a living thing, whether animal or vegetable, set apart to do a certain work, as the heart for pumping blood, the lungs for breathing, or the stem and leaves of a plant for conveying and digesting the sap. By "function" is meant the work or office that an organ has to perform.

9. The Cell. — In its strictly scientific sense this word is applied to the smallest portions of organized matter that go to make up a living body, whether vegetable or animal. It usually consists of a tiny membranous sac lined with a living semifluid substance called *protoplasm*, which ordinarily has one portion of denser consistency than the rest, called the *nucleus*. Within the protoplasmic lining are contained various watery fluids known as cell sap. These little sacs are packed together to build up the vegetable or animal structure as bricks are in building a wall. They are of various sizes and shapes. The containing membrane is called the

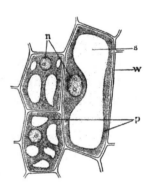

1. — Typical cells: *n*, nucleus; *p*, protoplasm; *w*, cell wall; *s*, sap.

cell wall. Cells can exist, however, without any wall, as mere specks or globules of protoplasm, but these are not common in vegetable structures. The essential part of every cell is the protoplasm with its nucleus. This substance, so far as we know at present, constitutes the physical basis of all life, and if the protoplasm loses its vitality, the cell dies and can no longer perform its functions of absorbing and retaining liquids. Slice a fresh beet in a vessel of water and a boiled one in another; how is the liquid affected in each? Account for the difference.

The name "cell" is also applied to the compartments into which the fruits and seed vessels of many plants are divided. This double meaning of an important term is unfortunate, but the context will always show in which sense it is to be taken, so that no confusion need result.

10. Tissue is a term used to denote any animal or vegetable substance that is composed of a particular kind of material and that performs a particular office or function. Thus, for instance, we have bony tissue and muscular tissue in animals; that is, tissue made of bone substance and of muscle substance and doing the work of bone and muscle respectively. So in plants, we have woody tissue, or tissue made of woody substance, and vascular tissue, or tissue made up of little conducting vessels, which have their especial functions to perform.

11. Appliances needed for General Use. — The only appliances necessary for the study of this book, besides the material furnished by the woods and fields about us, are so few and simple that there can be no difficulty in providing them. The following list comprises about all that are essential : —

Half a dozen glass jars; preserve jars or wide-mouthed bottles will answer.

Half a dozen soup plates or other shallow dishes for germinators.

Some good-sized bits of window glass for covering jars and dishes.

A garden trowel.

A good hatchet for use when the study of timber is taken up.

A very sharp knife — a razor is better, if it can be obtained — for making sections.

A small whetstone for sharpening knives.

A vial of tincture of iodine.

A pint of red ink; or, if preferred, a good coloring fluid can be made by purchasing an ounce or two of eosin from the druggist and mixing it with water.

A pot of photograph paste.

If a yard or two of India rubber tubing, a common bulb thermometer, and a pair of druggist's scales are added to the above list, the number of experiments that can be performed will be considerably increased.

12. Appliances for Individual Use. — In addition to the general outfit for the school, each pupil should be provided with —

A good. penknife.

A drawing book (or drawing paper) and a blank book for taking notes.

A book for dried specimens, made by sewing together two or three sheets of unsized paper, such as newspapers are printed on ; this can be purchased from a printer.

Two well-pointed pencils, one hard, the other medium.

A pair of dissecting needles ; wax-headed steel pins will do, but better ones can be made by running the heads of ordinary sewing. needles into handles of soft wood and gluing them in.

Two bits of glass, not larger than a visiting card, as thin and clear as can be obtained, for inclosing specimens that must be held up to the light for examination. The glass plates sold for photograph negatives serve well for this purpose.

A good hand lens. The glasses known as " linen testers " can be purchased for twenty-five cents apiece, and make very good magnifiers.

A special place ought to be provided in the schoolroom for storing all these articles, and the strictest order exacted in the care of them. They should always be ready when wanted, and never used for any other purpose.

13. Living Material. — A number of potted plants should always be kept in the schoolroom, especially in cities, for observation and experiment. Among those recommended for this purpose are the following : —

One or two ferns.

A calla lily, or other arum.

A young India rubber tree (*Ficus elastica*).

A pot of " wandering Jew " (*Zebrina pendula*). The plain, green-leaved varieties are best.

Some kind of prickly cactus. The common prickly pear (*Opuntia*) and the Mamillaria make good specimens.

A sedge; the umbrella plant (*Cyperus alternifolius*) or the Egyptian paper plant (*C. papyrus*), so common in greenhouses, will either of them do very well, though our native wild plants are always preferable when they can be obtained.

Healthy plants of oxalis and tropæolum.

A twining vine; hop, morning glory, kidney bean, etc.

A glass jar with one or two water plants, such as pondweed (*Potamogeton*), hornwort (*Ceratophyllum*), bladderwort (*Utricularia*), or pickerel weed (*Pontederia*), etc.

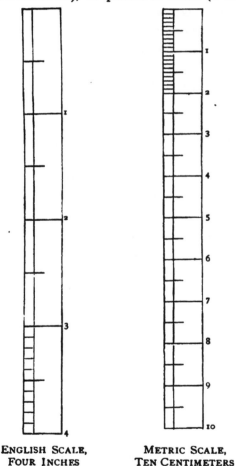

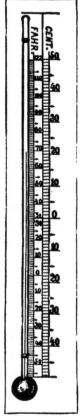

ENGLISH SCALE, FOUR INCHES

METRIC SCALE, TEN CENTIMETERS

2. — A comparative scale of the English and metric systems of length measure. One decimeter = 10 centimeters = 100 millimeters = approximately 4 inches.

3. — A comparative scale of the Centigrade and Fahrenheit thermometers. On the Centigrade scale 0 = the temperature of melting ice, and 100° = that of boiling water.

II. THE LEAF: ITS USES

TRANSPIRATION

MATERIAL. — Freshly cut sprigs of various kinds, bearing healthy leaves; a leaf of the white garden lily (*L. candidum*) or of the wandering Jew (*Zebrina pendula*); two hermetically sealing preserve jars; a little beeswax or tin foil; a bit of looking glass; a number of empty bottles with perforated stoppers or rubber cloth covers.

NOTE. — In order to avoid cumbering the pages of the text with technical nomenclature, botanical names of specimens mentioned will be given only: First, in the case of foreign or little known species; Second, where the popular name is local or provincial, or where the same term is applied to several different plants; and Third, where special accuracy of designation is required.

14. Why Leaves wither. — Dry two self-sealing jars thoroughly, by holding them over a stove or a lighted lamp for a short time to prevent their "sweating." Place in one a freshly cut leafy sprig of any kind, leaving the other empty. Seal both jars and set them in the shade. Place beside them, but without covering of any kind, a twig similar to the one in the jar. Both twigs should have been cut at the same time, and their cut ends covered with wax or vaseline, to prevent access of air. At the end of six or eight hours look to see if there is any moisture deposited on the inside of either jar. If there is none, set them both in a refrigerator or other cool place, for half an hour, and then examine them again. On which jar is there a greater deposit of dew? How do you account for it? Take the twig out of the jar and compare its leaves with those of the one left outside; which have withered most, and why?

15. Transpiration. — We learn from experiments like the foregoing that one office of leaves is *transpiration*, or the giving off of moisture, just as animals do through the pores of the skin.[1] Now, we all know what happens to us if the perspiration glands of our body get stopped up, and hence we need not be surprised if hedgerows can not be kept vigorous and healthy by dusty roadsides, nor if even sturdy trees and shrubs take on a sickly look when the summer rain delays too long to give them their accustomed bath.

16. Stomata. — The transpiration pores of leaves are called *stomata* (sing. *stoma*) from a Greek word meaning "mouths." Generally they are too small to be seen without a compound microscope, but their presence can be made manifest by a simple experiment. Place a bit of looking glass against your cheek or your arm on a warm day, and it will soon be covered with a film of moisture from the skin. Next, place the glass in contact with the under side of a healthy growing leaf for thirty to forty-five minutes, and see if you can detect any moisture on it. The deposit will probably be fainter than that from the skin, but the presence of any at all will show that the leaf transpires.

There are a few plants, such as the white lily of the gardens (*L. candidum*) and the wandering Jew, in which the stomata are large enough to be seen with a hand lens. The common iris also shows them, though not so distinctly. Strip off from the under side of such a leaf a portion of the epidermis, or outer covering. Place it

4. — Portion of the epidermis of the garden balsam, highly magnified, showing the very sinuous walled epidermis cells and three stomata (*after* GRAY).

5, 6. — Stomata of white lily leaf: 5, closed; 6, open (GRAY).

between two bits of glass with the outside uppermost, and

---

[1] Transpiration, though similar in external effects to the perspiration of animals, must not be confounded with it, as the two functions are physiologically quite different.

examine it with a good lens. Hundreds of little eye-shaped dots will be seen covering the surface, which can easily be recognized, by comparison with the accompanying Figures, as stomata. Examine a portion of the epidermis from the upper side of the leaf; are the stomata distributed equally on both sides, and if not, on which are they thickest?

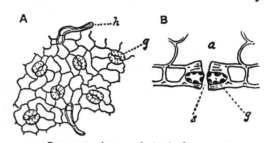

7.—Stomata of an oak leaf: A, a small piece (highly magnified) with under epidermis removed to show stomata, *g*, and minute hairs, *h*. B, a stoma in vertical median section, cut across its longer axis; *a*, intercellular space; *g*, guard cell; *s*, orifice of stoma.

Which side of the epidermis seems to be most active in the work of transpiration?

17. Distribution of Stomata. — While stomata are generally most abundant on the under side of leaves, where they are protected from excessive light and heat, this is not always the case. Similar openings occur also on young stems, and are called *lenticels*. In vertical leaves, like those of the iris, which have both sides equally exposed to the sun, stomata are distributed equally on both sides. In plants like the water lily, where the under surface lies upon the water, making transpiration in that direction impossible, they occur only on the upper side. Succulent leaves, as a general thing, have very few, because they need to conserve all their moisture. Submerged leaves have none at all; can you tell why?

18. Protection of Stomata. — In addition to their function of transpiration, stomata permit the entrance to the interior of the plant of atmospheric air containing carbon dioxide, a gaseous substance used by them in the formation of food. If they become choked up with water or other obstruction, the leaves can neither exhale their superfluous moisture nor take in air; hence these pores are protected by hairs, wax, and other water-shedding appendages. Plunge a sprig of the dwarf St. John's-wort (*Hypericum mutilum*) or of wandering Jew into water and notice the silvery appearance of the leaves, especially on the under side. In the iris it is the same on both sides; why? Remove the sprig from the water, and the leaves will be perfectly dry. In the wandering Jew, as may be seen with a good hand lens, this is due to the air imprisoned by little membranous appendages which surround the stomata and prevent the water from entering. In other cases, as cabbage, hypericum, etc., a coating of wax protects the transpiration pores, and it is the reflection of the light from the air entangled in these protective coverings that gives the leaves their silvery appearance under water.

19. Amount of Transpiration. — Few people have any idea of the enormous amount of water given off by leaves. It has been calculated[1] that an oak may have 700,000 leaves and that 111,225 kilograms of water (about 244,695 lbs.) may pass from its surface in the five active months from June to October, and 226 times its own weight of water may pass through it in a year. If this seems an extravagant estimate, we can easily make one for ourselves.

Fill three bottles with water, and cover them tightly with rubber cloth to prevent evaporation. Mark the point at which the water stands in the bottles, make a small puncture through the covers, and insert into one bottle the end of a healthy twig of peach or cherry, into the second a twig of catalpa, grape, or any other large-leaved plant, and into the third, one of magnolia, holly, or other thick, tough-leaved evergreen, letting the stems of all reach down well into the water. Care must be taken to select twigs of the same age, as the absorbent properties of very young stems are more injured by cutting and exposure than those of older ones. All the specimens should be cut under water if possible, as even an instant's exposure to the air will greatly diminish the activity of the cut surface. Peach

[1] See Marshall Ward, "The Oak."

is an excellent plant to experiment with, as its woody twigs
are not greatly affected by cutting, and it absorbs water
almost as rapidly as it transpires. At the end of twenty-
four hours note the quantity of liquid that has disappeared
from each glass. This will represent approximately the
amount absorbed by the leaves from the twigs to replace
that lost by transpiration. Which twig has transpired
most? Which least? Note the condition of the leaves
on the different twigs; have they all absorbed water as
rapidly as they have lost it? How do you know this?
Pluck the leaves from each twig, one by one, lay them on
a flat surface that has been previously measured off by the
aid of a rule, into a square of about thirty centimeters
(twelve inches) to a side, containing nine hundred square
centimeters (one hundred forty-four square inches), and
thus form a rough estimate of the area covered by each
specimen. Measure the amount of water transpired by
filling up each bottle to the original level, from a common
medicine glass, or if this cannot be obtained, use a table
spoon, counting two spoonfuls to the ounce. Make the
best estimate you can of the number of leaves on each tree,
and calculate the number of kilograms (or pounds) of water
it would give off at that rate in a day. In one experiment
a peach twig containing thirty-one leaves gave off three-
quarters of an ounce of water in twenty-four hours; how
many pounds would that be for the tree, estimating it to
bear eighteen thousand leaves? As the tissues of a growing plant are much more active than those of a severed
branch, calculations of this kind are not likely to exceed
the truth, even when we take into consideration the fact
that the twig in the experiment has unlimited water, which
the roots of a growing plant have not always.

These experiments may be varied at the option of the
teacher as time and opportunity may permit, so as to test
the absorbing and transpiring properties of any number of
plants or of the same plant at different stages of growth.
They will succeed best in dry, warm weather, as the work
of transpiration is then most active.

20. Practical Effects of Transpiration. — Where does all
this moisture come from? If the water in the last experiment is colored with a little eosin or with red ink, its
course can be traced through the stem into the leaves. In
growing plants the earth takes the place of our tumbler of
water, and from it the moisture is drawn up by the roots
and conveyed through the stem to the leaves. Thus we see
that trees are constantly acting as great pumps, drawing
up water from the lower strata of the soil and distributing
it to the thirsty air in summer. As the water given off
by transpiration is in the form of vapor, it must draw from
the plant the amount of heat necessary for its vaporization,
and hence it has the effect of making the leaves and the air
in contact with them cooler than the surrounding medium.

21. The Cause of Transpiration. — The reason why
plants exhale such large quantities of water is because they
get part of their food from mineral and other substances
dissolved in the water of the soil, but this food is in such a
diluted state that enormous quantities of the liquid containing it must be taken up in order to give the plant the nourishment it requires. This liquid travels through the stem
as sap, and after all the food substance has been extracted,
the waste water is exhaled by the leaves. Sometimes the
roots absorb moisture faster than the leaves can transpire
it ; the water then exudes through the stomata and settles
in drops on the blade, causing the leaves to sweat, just
as our bodies do under similar conditions. Sometimes, on
the other hand, the leaves transpire faster than the roots
can absorb, and then the plant wilts.

PRACTICAL QUESTIONS

1. Do you see any connection between the facts just stated and the
stories of "weeping trees" and "rain trees" that we sometimes read
about in the papers ? (Section 21.)

2. Can you explain the fact sometimes noticed by farmers, that in
wooded districts, springs which have failed or run low during a dry
spell sometimes begin to flow again in autumn when the trees drop
their leaves, even though there has been no rain? (19, 20.)

3. Other things being equal, which would have the cooler, pleasanter atmosphere in summer, a well-wooded region or a treeless one? (20.)

4. Could you keep a bouquet fresh by giving it plenty of fresh air? (14.)

5. Why does a withered leaf become soft and flabby, and a dried one hard and brittle? (9, 14.)

6. Why do large-leaved plants, as a general thing, wither more quickly than those with small leaves? (14–19.)

7. Is the amount of water absorbed always a correct indication of the amount transpired? Explain. (20, 21.)

8. Why must the leaves of house plants be washed occasionally to keep them healthy? (15–18.)

9. Why is it so hard to get trees to live in a large manufacturing town? (15, 18.)

RESPIRATION AND FOOD PRODUCTION

MATERIAL. — A green aquatic plant of some kind in a glass of water; two wide-mouthed glass jars; a bent glass or rubber tube, and a shallow dish of water; boiled bean or tropæolum, or other green leaves; a half pint of alcohol; some tincture of iodine; a strip or two of tin foil.

22. Leaves give off Oxygen. — Place in a glass of water a green aquatic plant of any kind; the common brook silk (*Spirogyra*) found in almost every pool will answer. Set it in the sunlight and place beside it another similar vessel containing nothing but water, and also a third vessel containing a piece of the same plant immersed in water from which the air has been expelled by boiling. After a time bubbles will be seen rising from the first vessel. Air bubbles will usually form on the bottom and sides also, but these are caused by the expansion of gases contained in the liquid, as will be evident on comparing them with similar phenomena in the jar containing only water, and must not be confounded with the gas given off by the plant. Remove the vessel from the light, and the bubbles will soon cease to appear, but will begin to form again if restored to the sunshine, thus showing that their production can take place only in the light. Do any bubbles at all appear in the glass with the boiled water?

It has been proved by chemical analysis that these bubbles are oxygen, which the plant has been separating from the gases mixed with the water, and giving off. It is even more active in separating oxygen from the air, but the process is not visible to the eye, because we cannot see a gas except in the form of bubbles. Water is used not as an aid to the plant in the performance of its function, but in order to enable us to see the result.

23. Leaves as Purifiers of the Atmosphere. — Fill two tumblers with water, to expel the air, and invert in a shallow dish of water, having first introduced a freshly cut sprig of some healthy green plant into one of them.

8. — Experiment for showing how leaves purify the atmosphere.

Then by means of a bent tube blow into the mouth of each tumbler till all the water is expelled by the impure air from the lungs. Set the dish in the sunshine and leave it, taking care that the end of the cutting is in the water of the dish. After forty-eight hours remove the tumblers by running under the mouth of each, before lifting from the dish, a piece of glass well coated with vaseline (lard will answer) and pressing it down tight so that no air can enter. Place the tumblers in an upright position, keeping them securely covered. Fasten a lighted taper or match to the end of a wire, plunge it quickly first into one tumbler, then into the other, and note the result. It is an established fact that a light will not burn in an impure atmosphere; this is why well cleaners send down a lighted candle before going into a well themselves. What are we to infer from the effects observed as to the action of the plant upon the atmosphere?

This experiment will not succeed unless performed very carefully, and the air must be absolutely excluded from the tumblers until the instant the taper is plunged in.

24. Leaves as Food Makers. — It thus appears that plants are constantly reversing the effects of animal respiration by giving off oxygen and absorbing carbon dioxide from the air. Besides acting as digestive and assimilating organs, leaves are the laboratories in which plant food is manufactured out of the crude materials brought up from the soil by the sap, and those absorbed through the stomata from the gases of the atmosphere. Carbon dioxide taken from the atmosphere is somehow used up in this operation, and the oxygen, which is not needed by the plant, is given back to be consumed by animals. This is the most important work the leaf has to do, and because it can take place only in the light, has been named by botanists *Photosynthesis*, a word which means "building up by means of light," just as *photography* means "drawing or engraving by means of light."

25. Why Leaves are Green. — Has the color of the leaf anything to do with this function? It will help to a correct answer if we remember that herbs grown in the dark, and parasites like the dodder and Indian pipe (*Monotropa*), which steal their food ready made from the tissues of other plants, and so have no need to manufacture it for themselves, always lose their green color. Place a seedling of oats or other rapidly growing shoot in the dark for a few days and note its loss of color. Leave it in the dark indefinitely, and it will lose all color and die. Hence we may conclude that there is some intimate connection between the action of light and the green coloring matter of leaves. This green matter is called *Chlorophyll*, a word meaning "leaf green," and physiologists tell us that through its agency the crude substances brought up from the soil in the sap and the carbon dioxide of the air are converted into nourishment.

26. Starch as Plant Food. — It is the office of chlorophyll to manufacture a particular class of plant foods known as *carbohydrates*. The commonest and most important of these is starch, the presence of which can generally be detected without much difficulty. Boil a few leaves of bean or sunflower, tropæolum, etc., for about fifteen minutes, and soak them in alcohol until all the chlorophyll is dissolved out. Rinse them in water, and soak the leaves thus treated, in a weak solution of iodine for half an hour; then wash them and hold them up to the light. Iodine turns starch blue; hence if there are any blue spots on the leaves, what are you to conclude? Other food substances can be detected by proper tests, but none of them so readily as starch.

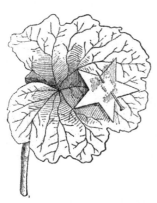

9. — Leaf arranged with a disk of tin foil to exclude light from a portion of the surface.

27. Necessity of Light and Air. — Exclude the light from parts of healthy leaves on a growing plant of tropæolum, bean, etc., by placing bands or patches of tin foil over them. Leave in a bright window, or preferably out of doors, for twenty-four to forty-eight hours, and then test for starch as in the last experiment; do you find any in the shaded spots?

Cover the lower side of several leaves with vaseline or other oily substance so as to exclude the air, and after a day or two test as before.

From these experiments we learn that leaves can not do their work without light and air. The particular element of the atmosphere used by them in the process of food making is carbon dioxide, a poisonous gas that is being constantly produced by the decay of vegetable and animal matter, by the respiration of animals, and by combustion of all sorts. It constitutes about one fourth of one per cent of our atmosphere, and when the proportion rises much above this, the air becomes unfit to breathe, so that the work of plants in eliminating it is a very important one.

28. Respiration. — The leaf is also an organ of respiration; that is, it is always taking in oxygen and giving off carbon dioxide, just as animals do, but in such small quantities that the process is entirely obscured during the day by the much more active function of photosynthesis, or food making, which goes on at the same time. For this reason it was formerly believed that respiration, or the absorption of oxygen by plants, took place only at night, and some people were led to imagine from this that it is unwholesome to have potted plants in a bedroom; but the quantity of oxygen absorbed by green plants is so small as to be scarcely appreciable.

While the leaf is the principal organ of respiration, this function is carried on in other parts of the plant also, else it could not survive during the leafless months of winter. It goes on at all times, in all living parts, and the other leaf functions also are carried on, to some extent, in all green tissues.

29. Relation of Respiration to Other Functions. — The functions of photosynthesis and respiration are mutually complementary and interdependent, the one manufacturing food, and the other using it up, or rather marking the activity of those life processes by which it is used up. In this respect it is strictly analogous to the respiration of animals. The more we exert ourselves and the more vital force we expend, the harder we breathe, and hence respiration is more active in children than in older persons, and in working people than in those at rest. It is just the same with plants; respiration is always most energetic in germinating seedlings and young leaves, in buds and flowers, where active work is going on; hence such organs consume proportionately large quantities of oxygen and liberate correspondingly large quantities of carbon dioxide.

Fill a glass jar of two liters' capacity (about two quarts) with germinating seeds, or with flower buds or unfolding leaf buds arranged in layers alternating with damp cotton batting or blotting paper; close it tightly and leave it for twelve to twenty-four hours. If the jar is then opened and a lighted taper

10. — Arrangement of apparatus to show that carbon dioxide is given off by growing seedlings.

plunged in, it will be extinguished as quickly as in the empty tumbler in the experiment described in Section 23, thus showing that the process of respiration is more active in this case than the opposite function of taking in carbon dioxide and liberating oxygen. Insert a thermometer bulb and note the difference in temperature. In some of the arums, — calla lily, Jack-in-the-pulpit, elephant's ear (*Colocasia*), etc., — where a large number of small flowers are brought together within the protecting spathe, the rise of temperature is sometimes so marked that it may be perceived by placing a flower against the cheek.[1]

30. Metabolism. — The total of all the life processes of plants, including growth, waste, repair, etc., is summed up by botanists under the general term *Metabolism*. It is a constructive or building-up process when it results in the making of new tissues out of the food absorbed from the earth and air, and consequent increase of the plant in size or numbers. But, as in the case of animals, so with plants, not all the food provided is converted into new tissue, a part being decomposed and excreted as waste. In this sense, metabolism is said to be *destructive*, and, like other destructive processes (combustion, for instance), is always accompanied by the liberation of energy, — heat, as we have seen, being an invariable accompaniment. The waste in healthy plants is always, of course, less than the gain, and a large portion of the food material is in all cases laid by as a reserve store. For this reason, photosynthesis, which is a constructive process, is usually more energetic than respiration, which is the measure of the destructive change of materials that attends all life processes.

It is evident also, from what has been said, that growth and repair of tissues can take place only so long as the plant has abundant oxygen for respiration, since the food material manufactured by it must be decomposed into the

[1] See Sachs, " Physiology of Plants."

various substances required by the different tissues before it can be appropriated by them.

PRACTICAL QUESTIONS

1. Why do gardeners bank up celery to bleach it? (25.)

2. Why are the buds that sprout on potatoes in the cellar white? (25.)

3. Why does young cotton look so pale and sickly in long-continued wet or cloudy weather? (25.)

4. Why do parasitic plants generally have either no leaves or very small, scalelike ones? (25.)

5. The mistletoe is an exception to this; can you tell why? (184.)

6. Could an ordinary self-supporting plant live without green leaves? (26, 27.)

7. Are abundance and color of foliage any indication of the health of a plant? (24, 26.)

8. Is the practice of lopping and pruning very closely, as in the process called "pollarding," beneficial to a tree under ordinary conditions? (18, 21, 24, 26.)

9. Why is it wise to trim a tree close when we transplant it? (20, 21.)

10. Why should transplanting be done in winter or very early spring, when the leaves are off? (19, 20.)

11. Name some plants of your neighborhood that grow well in the shade.

12. Compare in this respect Bermuda grass and Kentucky blue grass; cotton and maize; horse nettle (*Solanum carolinense*) and dandelion; beech, oak, red maple, dogwood, pine, cedar, holly, magnolia, etc.

13. Why are evergreens more abundant in cold than in warm climates? (19, exp.)

14. Is it wholesome to keep blooming plants in a bedroom? Leafy ones?

15. Why, in each case? (23, 28.)

THE TYPICAL LEAF AND ITS PARTS

MATERIAL. — Leaves of as many different kinds as can conveniently be obtained, showing their various modes of attachment, shapes, texture, etc. For stipules, leaves on very young twigs should be sought for, as these bodies often fall away soon after the leaves expand. The rose, Japan quince (*Pyrus japonica*), willow, strawberry, pansy, pea, and young leaves of apple, peach, elm, oak, beech, tulip tree (*Liriodendron*), India rubber tree (*Ficus elastica*), magnolia, etc., furnish good examples of stipules.

31. Parts of the Leaf. — Examine a young, healthy leaf of apple, quince, elm, etc., as it stands upon the stem, and

notice that it consists of three parts: a broad expansion called the *blade;* a leaf stalk or *petiole* that attaches it to the stem; and two little leaflike, or bristlelike bodies at the base, known as *stipules.* Make a sketch of any leaf provided with all these parts and label them respectively blade, petiole, and stipules.

11. — A typical leaf and its parts: *b*, blade; *p*, petiole; *s, s*, stipules.

32. Stipules. — These three parts make up a perfect or typical leaf, but as a matter of fact, one or more of them is usually wanting. The office of stipules, when present, is generally to subserve in some way the purposes of protection. In many cases, as the fig, elm, beech, oak, magnolia, etc., they appear only as protective scales that cover the bud during winter, and fall away as soon as the leaf expands. When *persistent,* that is, enduring, they sometimes take the form of spines and thorns, as in the black locust and spiny clotbur (*Xanthium spinosum*). The sheathing stipules of the smartweeds and bindweeds (*Polygonum*) serve to strengthen the stem at the joints (Fig. 13), and the adnate stipules (Fig. 14) of the rose, clover, strawberry, etc., may serve either as water holders or as shields against climbing insects. In the smilax and some other vines they appear as tendrils for climbing, while in other cases, as the garden pea and pansy, they become large and leaflike, or may even usurp the place of

12. — Spiny stipules of Clotbur.

13. — Sheathing stipules of "prince's feather" (*Polygonum orientale*) (GRAY).

the leaves altogether, as in the *Lathyrus aphaca* (Fig. 17),

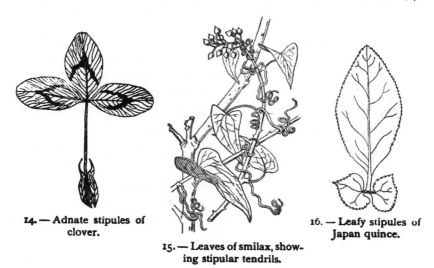

14. — Adnate stipules of clover.

15. — Leaves of smilax, showing stipular tendrils.

16. — Leafy stipules of Japan quince.

a near relative of the sweet pea, where they function as foliage. But under whatever form they occur, their true nature may be recognized by their position on each side of the base of the petiole, and not in the *axil*, or angle formed by the leaf with the stem.

17. — Leaf of *Lathyrus aphaca*, reduced to a pair of stipules and a tendril (*after* GRAY).

33. Petioles. — The normal use of the petiole is to secure a better light exposure for the leaves, but like other parts of the leaf, it is subject to modifications. In some vines, such as the jasmine nightshade and tropæolum of the gardens, it is twisted into a tendril for climbing. Occasionally the leaf blade disappears altogether and the leaf stalk takes its place, as in some of the Australian acacias frequently seen ·in greenhouses. · Simulated leaves of this kind can generally be distinguished by their edgewise position, the blades of true leaves being usually horizontal. Other instances occur, such as the onion, jonquil, hyacinth, etc., where the distinction, if any exists, is difficult to make out.

In the sycamore, the base of the petiole is hollowed out into a socket to protect the bud of the season (Fig. 20).

34. Leaf Attachment. — When the petiole is wanting altogether, as is often the case, leaves are said to be *sessile*, that is, *seated* on the stem, and their bases are described by various terms suggestive of the mode of attachment. You can frame your own definition of these terms by an inspection of the accompanying figures, or better still, of some of the sample plants named in connection with each.

Clasping (Fig. 21): Wild lettuce (*Lactuca*), chicory, sow thistle (*Sonchus*), poppy, stem leaves of turnip, mustard, etc.

Decurrent (Fig. 22): Thistle, sneezeweed (*Helenium autumnale*), comfrey (*Symphytum*).

Connate (Fig. 23): The upper leaves of boneset (*Eupatorium perfoliatum*) and trumpet honeysuckle (*Lonicera sempervirens*).

Perfoliate (Fig. 24): Bellwort (*Uvularia perfoliata*).

Peltate, or shield-shaped (Fig. 25): Castor oil plant, tropæolum, May apple (*Podophyllum*), water pennywort (*Hydrocotyle*).

Equitant (Fig. 26): Iris, sweet flag (*Acorus calamus*), blackberry lily (*Belamcanda chinensis*).

35. The Use of Botanical Language. — These terms and those which follow are not to be learned by heart, but are given here merely for convenience of reference. Botanists have invented a number of useful terms for describing things briefly and accurately, and while they are not to be regarded as of any importance in themselves, it is impossible to get along without some knowledge of them ; for besides furnishing a sort of universal vocabulary, intelligible to botanists everywhere, they enable us to say in two or three words what it would otherwise require as many lines or perhaps paragraphs to express. In other words, they are a sort of labor-saving device which every botanist must learn how to use, as no good workman can afford to be ignorant of the tools of his profession.

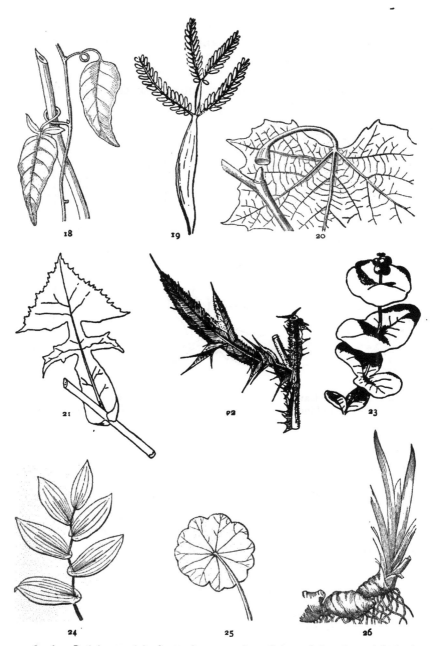

18–26. — Petioles, and leaf attachment: 18, petioles of jasmine nightshade (*Solanum jasminoides*) acting as tendrils; 19, acacia, showing petiole transformed to leaf blade; 20, petiole of sycamore hollowed out to protect the bud of the season; 21, clasping leaf of lactuca; 22, decurrent leaf of thistle; 23, connate leaves of honeysuckle; 24, perfoliate leaves of uvularia; 25, peltate leaf of tropæolum; 26, equitant leaves of iris. (18, 20, 23, 24, 25, and 26, *after* GRAY.)

36. Shape and Texture of Leaves. — Examine a number of leaves of different kinds and see how they differ from each other in regard to —

General Outline: whether round, oval, heart-shaped, lanceolate, etc. (Figs. 27–33).

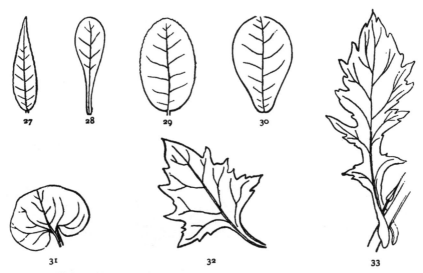

27–33. — Shapes of leaves: 27, lanceolate; 28, spatulate; 29, oval; 30, obovate; 31, reniform, or kidney-shaped; 32, deltoid; 33, lyrate. (27–31, *after* GRAY.)

Base: tapering, obtuse, truncate, cordate, etc. (Figs. 34–38).

34–38. — Bases of leaves: 34, cordate; 35, sagittate; 36, oblique; 37, auricled; 38, hastate.

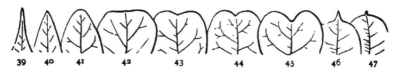

39–47. — Apexes of leaves: 39, acuminate; 40, acute; 41, obtuse; 42, truncate; 43, 44, emarginate; 45, obcordate; 46, cuspidate; 47, mucronate (GRAY).

Apex: acute, acuminate, emarginate, etc. (Figs. 39–47).

Margins: some being unbroken or *entire*, others variously toothed and cut (Figs. 48–53).

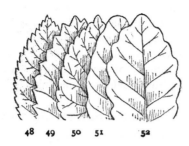

48–53. — Margins of leaves: 48, serrate; 49, dentate; 50, crenate; 51, undulate; 52, sinuate; 53, runcinate leaf of dandelion.   (48–52, *after* GRAY.)

Symmetry: that is, whether the two halves are alike, so that if folded over on each other they would coincide.

Texture: whether thick or thin, fleshy and soft, hard and brittle, or tough and leathery (*coriaceous*).

Surface: smooth and shining (*glabrous*); wrinkled (*rugose*); hairy (*pubescent*); covered with a bloom (*glaucous*); moist and sticky (*viscid*, or *glandular*).

PRACTICAL QUESTIONS

1. Tell the nature and use of the stipules in such of the following plants as you can find: tulip tree; fig; beech; apple; willow; pansy; garden pea; Japan quince (*Pyrus japonica*); sycamore; rose; paper mulberry (*Broussonetia*).

2. State what differences and resemblances you observe between the leaves of the elm, beech, birch, alder, hackberry, hornbeam.

Between the hickory, ash, common elder, walnut, ash-leaved maple (*Negundo*), ailanthus, sumac.

Between the persimmon, black gum, buckthorn, papaw (*Asimina*), sourwood (*Oxydendron arboreum*).

Between chinquapin, chestnut, and chestnut oak.

Any other sets of leaves may be substituted for those named, the object being merely to form the habit of distinguishing readily the differences and resemblances between leaves that bear some general likeness to one another.

Notice that the general resemblances are not confined to plants of closely related species: what other causes may influence them?

VEINING

MATERIAL. — A specimen of each of the different kinds of veining. For parallel veining any kind of arum, lily, or grass will do ; for net veining, ivy, maple, elm, or peach, etc. Classes in cities can use leaves from potted plants of wandering Jew (*Zebrina pendula*), calla lily, and other easily cultivated specimens, or blades of grass, plantain, and various parallel and net veined weeds can be picked up here and there, even in the largest cities. Have a number of leaves placed with their cut ends in red ink from three to six hours before the lesson begins.

54. — Parallel-veined leaf of lily of the valley (*after* GRAY).

37. Parallel and Net Veining. — Compare a leaf of the wandering Jew, garden lily, or any kind of grass, with one of cotton, maple, ivy, etc. Hold each up to the light, and note carefully the veins or little threads of woody substance that run through it. Make a drawing of each so as to show plainly the direction and manner of veining. Write under the first, *Parallel veined*, and under the second, *Net veined*. This distinction of leaves into parallel and net veined corresponds with another important difference in plants, existing in the seed, and is used by botanists in distinguishing the two great classes into which seed-bearing plants are divided.

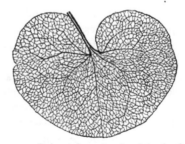

55. — Palmately net-veined leaf of *Asarum europæum.*

38. Pinnate and Palmate Veining. — Next, compare a leaf of the canna, or of any of our common garden arums, with one of the elm, peach, cherry, etc., or with a leaflet of the rose or clover. Hold both up to the light and observe carefully the veins and reticulations. What resemblance do you notice between the two? What difference? Which is parallel veined and which is net veined? Make a drawing of each, and compare with the first two. Notice that in the last, the petiole seems to be continued in a large central vein, called the *Midrib*, from which the secondary veins branch off

on either side just as the pinnæ of a feather do from the quill; whence such leaves are said to be *pinnately*, or *feather* veined. In the cotton, maple, ivy, etc.,

56. — Pinnately paral-lel-veined leaf of cala lily (*after* GRAY).

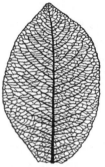

57. — Pinnately net-veined leaf of a willow.

on the other hand, the petiole breaks up at the base of the leaf (Fig. 55) into a number of primary veins or ribs, which radiate in all directions like the fingers from the palm of the hand; hence, such a leaf is said to be *palmately* veined.

39. Ribbed Leaves. — Net-veined leaves are sometimes ribbed in a way that might lead an inexperienced observer to confound them with parallel-veined ones. Compare, for instance, a leaf of the wild smilax (often improperly called bamboo), or of the common plantain, with one of the kind represented in Figure 54. A little inspection will show that in both the ribs all proceed from the same point at the top of the petiole, as in other leaves of the palmate kind, of which they are varieties, but the reticulations between the ribs in the smilax and plantain show that they belong to the net-veined division.

58. — Ribbed leaf of plantain.

40. Parallel-veined and Straight-veined Leaves. — In some pinnate leaves, like the elm, beech, birch, dogwood, etc., the secondary veins are so straight and regular that beginners are apt to confound them with the parallel kind represented in Figure 56, but this mistake need never occur if the reticulations of the smaller veinlets are noted. Then, too, it must be observed that in a pinnately parallel-veined leaf the secondary veins do not separate from the midrib in such sharp, clear-cut angles as we see in the beech and elm, but seem to flow into it and mingle gradually with it, so that the midrib has the appearance of being made up of the overlapping fibers of the smaller veins, as in Figure 56.

59. — Straight-veined leaf of dogwood.

41. Use of the Veins. — Hold up a stiff, firm leaf of any kind, like the magnolia, holly, or India rubber, to the light, having first scraped away a little of the under surface, and examine it with a lens. Compare it with one of softer texture, like the peach, maple, grape, cotton, clover, etc. In which are the veins closest and strongest? Which is most easily torn and wilted? Tear a blade of grass longitudinally and then crosswise; in which direction does it give way most readily? Tear apart gently a leaf of cotton, maple, or ivy, and one of elm or other pinnately-veined plant; in which direction does each give way with least resistance? What would you judge from these facts as to the office of the veins?

42. Effect upon Shape. — By comparing a number of leaves of each kind, it will be seen that the feather-veined ones tend to assume elongated outlines (Figs. 16, 33, 53), while the palmate veining produces more broad and rounded forms (Figs. 25, 55, 61). Notice also that the straight, unbroken venation of parallel-veined leaves is generally accompanied by smooth, unbroken margins, while the irregular, open meshes of net-veined leaves are favorable to breaks and indentations of all kinds.

43. Veins as Water Pipes. — Examine a leaf that has stood in red ink for two or three hours. Do you see evidence that it has absorbed any of the liquid? Cut across the blade and examine with a lens. What course has the absorbed liquid followed? What use does this indicate for the veins, besides the one already noted?

We thus see that the veining serves two important purposes in the economy of the leaf; first, as a skeleton, or framework, to support the expanded blade; and second, as a system of supply pipes, or waterworks for conveying the sap out of which its food is manufactured.

The microscope shows us that the veins are made up of clusters or bundles of woody fibers, mixed with long, tubular cells that serve as vessels for conducting the sap; hence they are called *fibrovascular* bundles; which means bundles composed of fibers and conducting vessels. In this way the veins get both their hardness and their water-conducting power. The tough, stringy threads that protrude from the petiole of a plaintain leaf when broken are made of fibrovascular bundles that supply the leaf blade.

PRACTICAL QUESTIONS

1. In selecting leaves for decorations that are to remain several hours without water, which should you prefer, and why: Smilax or Madeira vine (*Boussingaultia*)? Ivy or Virginia creeper? Magnolia or maple? Maidenhair or shield fern (*Aspidium*)? (41, 43.)

2. Should you select very young leaves, or more mature ones, and why?

3. Can you name any parallel-veined leaves that have their margins lobed, or indented in any way?

4. Which are most common, parallel-veined or net-veined leaves?

5. Why do the leaves of corn and other grains not shrivel lengthwise in withering, but roll inward from side to side? (41.)

6. Can you name any palmately-veined leaves in which the secondary veins are pinnate? Any pinnately-veined ones in which the secondary veins are palmate?

7. Account for the difference.

BRANCHED LEAVES

MATERIAL. — Lobed and compound leaves of various kinds. Many good examples can be found among the weeds growing on vacant lots in cities.

44. Lobing. — Compare the outline of a leaf of maple or sweet gum with one of oak or chrysanthemum. Do

you perceive any correspondence between the manner of
lobing or indentation of their margins, and the direction
of the veins? To what class would you refer each one?

The lobes themselves may be variously cut, as in the

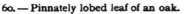

60. — Pinnately lobed leaf of an oak.   61. — Palmately lobed leaf of grape.

fennel and rose geranium, thus giving rise to twice-cleft,
thrice-cleft, four-cleft, or even still more intricately divided
leaves. Where the divisions are very deep it may sometimes be a little puzzling to decide whether they are not

62. — Pinnately divided leaf   63. — Palmately parted leaf of tall butter-
      of a buttercup.   cups.

separate leaflets, but if there is the merest thread of green
connecting the segments, as in Figures 62 and 63, it is considered a simple lobed leaf.

45. Compound Leaves. — Compare with the specimens
just examined a leaf of horse-chestnut, clover, or Virginia

creeper, etc., and one of rose, black locust, vetch, or other

pinnate leaf. Notice that each of these last is made up of entirely separate divisions or leaflets, thus

64. — Pinnately compound leaf of black locust.

65. — Palmately compound leaf of horse-chestnut.

forming a *compound leaf.* Notice also that the two kinds of compound leaves correspond to the two kinds of veining and lobing, so that we have palmately and pinnately compound ones. In pinnate leaves the continuation of the common petiole along which the leaflets are ranged is called the *Rhachis.*

46. Trifoliolate Leaves. — In a trifoliolate leaf, or one of three parts, it is often difficult for a beginner to decide whether the divisions are palmate or pinnate. To settle this question, compare a leaf of lucerne, beggar's ticks, or

66. — Pinnately trifoliolate leaf of a desmodium.

67. — Palmately trifoliolate leaf of wood sorrel.

bush clover (*Lespedeza*), with one of wood sorrel (*Oxalis*), or any common clover, and observe the mode of attachment of the terminal leaflet. When the common petiole is prolonged ever so little beyond the insertion of the two lateral leaflets, so as to form a rhachis, as in Figure 66,

the leaf is pinnately trifoliolate; but if all three appear to spring directly from the top of the petiole, as in Figure 67, it is palmate. A good example of a pinnately trifoliolate leaf, and one which it is important to learn and remember, is the poison ivy.

47. Unity of Plan in Nature. — Notice how the same plan of structure runs unchanged through all these variations. If an oak or a tansy leaf

68. — Poison ivy.

were cut through to the midrib, we should have a pinnately compound leaf, while a sweet gum or a maple cut in the same way would give rise to a palmately compound one.

48. The Branching of Leaves. — Lobed and compound leaves represent mere degrees of branching. Notice, however, that their mode of branching differs from that of stems in having the branches all in the same plane, like figures cut out of a single sheet of paper. This is what we should expect in the case of expanded bodies whose primary object is exposure to the light.

49. What makes a Compound Leaf. — Some botanists do not regard a branched leaf as compound unless the leaflets are jointed to the common petiole so that they break and fall away separately in autumn, like those of the ash, horse-chestnut, china tree, etc. According to this defi-

69. — Leaf of common orange.

70. — Leaf of trifoliolate orange.

nition, the single leaf of the orange and lemon is compound, for it is jointed to the petiole like those of the ash and hickory. This view is supported by the fact that some species of orange have trifoliolate leaves.

PRACTICAL QUESTIONS

1. State whether such of the following leaves as you can find are lobed or compound: cinquefoil, wood anemone, tree fern (*Polypodium incanum*), buttercups, Dutchman's breeches (*Dicentra*), mayweed, chamomile, yarrow, tickseed (*coreopsis*), shield fern, agrimony, tomato, tansy, cosmos, cypress vine, wild carrot, larkspur, strawberry, monkshood, celandine.

2. Which of the following are pinnately and which palmately trifoliolate?   Lucerne, red clover, Japan clover (*Lespedeza striata*), beggar's ticks (*Desmodium*), sweet clover (*Melilotus*), kidney bean, strawberry.

3. Name some of the favorite shade trees of your neighborhood; do they, as a general thing, have their leaves entire, or branched and compound?

4. Which of the following are the better shade trees, and why: pine. white oak, mimosa (*Albizzia*), sycamore, locust, horse-chestnut, fir. maple, linden, China tree. cedar, ash?

5. Which would shade your porch better, and why: cypress vine, grape, gourd, morning-glory, wistaria, clematis, smilax, kidney bean. Madeira vine, rose, yellow jasmine, passion flower?

PHYLLOTAXY, OR LEAF ARRANGEMENT

MATERIAL. — Twigs of any kinds of plants with opposite and alternate leaves. For the different orders of alternate arrangement, elm, ivy, basswood, wandering Jew, or any kind of grass will show the first: alder, birch, or any kind of sedge, the second; peach, oak, cherry, or almost any of our common trees and shrubs, the third (in cities, a potato plant grown in a pot may be used). In selecting specimens, choose straight, young twigs in order to avoid confusion from twisting of the stem that often occurs in older specimens on account of light exposure, or from other causes.

50. Alternate and Opposite Leaves. — Compare the arrangement of leaves on a twig of elm or basswood, or on a culm of grass, etc., with that of the foliage of the maple, lilac, or honeysuckle. Make a vertical diagram of each, as

shown in Figures 73 and 74, illustrating the two modes of arrangement. Label the point at which the leaf is in-

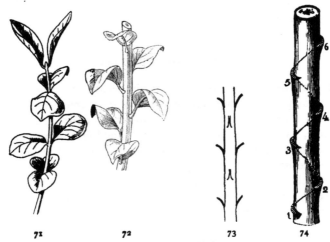

72 73 74

71–74.—Arrangement of leaves: 71, opposite-leaved twig of spindle tree; 72, alternate-leaved twig of apple; 73, vertical diagram of opposite-leaved twig; 74, vertical diagram of two-ranked twig of elm. (71, 72, *after* Gray.)

serted, the *node;* the space between any leaf and the one next above or below it, the *internode;* and angle between

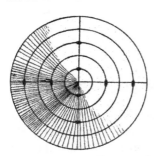

75.—Horizontal diagram of opposite leaves.

the leaf and the stem, where you see the bud, the *axil.* How many leaves are there at a node in the elm and basswood? How many in the maple and honeysuckle? Are the two consecutive pairs of leaves in the latter placed directly over each other, or at right angles? How far round from the first leaf does the second stand in the elm, grass, etc.? How does its position differ from that of the same leaf in the opposite mode of insertion? How many leaves must be passed in order to complete a turn round the stem, and what leaf in numerical order stands directly above the first? Draw a horizontal diagram of both twigs repre-

76.—Horizontal diagram of two-ranked leaves.

senting the two kinds of arrangement as viewed from above. Notice that if we join the leaves in the opposite arrangement by dotted lines we shall get a series of circles (Fig. 75), while the alternate arrangement will give a spiral (Fig. 76).

These two kinds of insertion, the alternate and opposite, represent the fundamental forms of leaf disposition. There may be varieties of each, but no matter what minor differences exist, all may be referred to one of these two modes.

51. Whorled and Fascicled Leaves. — Where more than two leaves occur at a node they constitute a *whorl*, or *verticel*, as in the trillium and common cleavers (*Gallium*). There is no limit to the number of leaves that may be in a whorl except the space around the stem to accommodate them.

A *fascicle*, or cluster, of which the pine and larch furnish examples, is composed of alternate leaves with very short internodes, which bring the leaves so close together as to give them the appearance of a whorl.

77-78. — Whorls and fascicles: 77, whorled leaves of Indian cucumber; 78, fascicled leaves of pine.

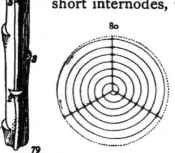

79-80. — Three-ranked arrangement: 79, vertical diagram; 80, horizontal diagram.

52. Varieties of Alternate Arrangement. — The kind of alternate arrangement just described is called the two-ranked, because it distributes the leaves in two rows on opposite sides of the stem; in other words, each is just halfway round from the one next above or below it. Other common forms of the alternate or spiral arrangement are the three-ranked (Figs. 79 and 80), in which three leaves are passed in completing a turn round the stem, the fourth in vertical order standing over the first; and the five-ranked (Figs. 81 and 82), in which five leaves are passed in making *two* turns, and the sixth in numerical order stands above the first. This is the commonest of all the modes of insertion, and the one that prevails among our forest trees and shrubs. The two-ranked is characteristic of the grass family, and the three-ranked of the sedges, though both occur among other plants as well. Specimens of all the kinds mentioned should be examined and compared with the diagrams. There are other and more complicated arrangements, but they are not common enough to demand attention here.

81–82.—Five-ranked arrangement: 81, vertical diagram; 82, horizontal diagram.

53. Relation between Phyllotaxy and the Shape of Leaves. — Compare the vertical distance between leaves on the same and on different twigs; are the internodes all of the same length? Where the internodes are short, the leaves will be crowded together in closer vertical rows. A compact arrangement of this sort tends to shut off light from the lower leaves; hence, in plants where it prevails, the leaves are apt to be long and narrow in proportion to the frequency of the vertical rows. The yucca, oleander, Canada fleabane, and bitterweed (*Helenium tenuifolium*), all illustrate this law.

83.— Narrow leaves in crowded vertical rows.

On the other hand, where the internodes are long or the vertical rows few, the leaves tend to assume more broad and rounded shapes, as in the cotton, hollyhock, sunflower, etc. If the blades are much cut and lobed, so that the

84.—Long internodes and large leaves.

85.—Dissected leaves overlapping one another without injurious shading.

sun easily strikes through, they can bunch themselves in almost any way without injurious shading. The length of the internodes depends, to a large extent, upon the rapidity of growth, being usually much greater in vigorous young shoots and the terminal portion of the main stem than in the lateral branches.

PRACTICAL QUESTIONS

1. Strip the leaves from a twig of one order of arrangement and replace them with foliage from a twig of a different order; for instance, place basswood upon white oak, birch upon lilac, elm upon pear, honeysuckle on barberry, etc. Is the same amount of surface exposed as in the natural order?

2. What disadvantage would it be to a plant if the leaves were arranged so that they stood directly over one another? (24, 25, 27.)

3. Why are the internodes of vigorous young shoots, or scions, generally so long? (53.)

4. If the upward growth of a stem or branch is stopped by pruning, what effect is produced upon the parts below, and why?

5. Why does corn grow so small and stunted when sown broadcast for forage?

6. What is the use of "chopping" cotton?

LEAF ADJUSTMENT

MATERIAL. — Upright and horizontal twigs from the same plant, any kind obtainable. A potted plant of oxalis, spotted medick, white clover or Japan clover (*Lespedeza striata*), or any other irritable kind.

54. Leaves adjust themselves to Light. — Take two sprigs, one upright, the other horizontal, from any convenient shrub or tree — those with opposite, or two-ranked leaves, like the elm and linden will generally show this peculiarity best — and notice the difference in the position of the leaves. Examine their points of attachment and see how this is brought about, whether by a twist of the petiole or of the base of the leaf blades, or by a half twist of the stem between two consecutive leaves, or by some other

86, 87. — Adjustment of leaves to different positions: 86, upright; 87, procumbent.

means. Observe both branches in their natural position; what part of the leaf is turned upward, the edge or the surface of the blade? Change the position of the two sprigs, placing the vertically growing one horizontal, and the horizontal one vertical. What part of the leaves is turned upward in each? One need only glance at the sky on any bright day and see how the light falls to understand the meaning of this adjustment. Would the same amount of light and air be secured by any other?

Rose bushes and a few other plants sometimes take on a second growth in late summer and autumn. If you can find such a plant, bend a young vertical branch into a horizontal position, and a horizontal one into an upright position and fasten them there. Examine at intervals and note the adjustment of the new leaves as they develop.

55. Mosaics and Rosettes. — A very little observation will show that trees with horizontal or drooping branches, like the elm and beech, and vines growing along walls or trailing on the ground, generally display their foliage in flat, spreading layers, each leaf fitting in between the interstices of the others like the stones in a mosaic, whence this has been called the *mosaic* arrangement. In plants of more upright or bunchy habit, on the other hand, the leaves grow at

88. — Leaf mosaic of elm.

all angles, with a general tendency to cluster in rosettes at the end of the branches, as in the magnolia, horse-chestnut, sweet gum, etc., thus giving rise to what is known as the *rosette* arrangement.

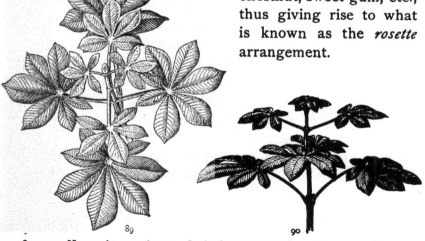

89, 90. — Horse-chestnut leaves: 89, leaf rosette seen from above; 90, the same seen sidewise, showing the formation of rosettes by the lengthening of the lower petioles.

56. Leaf Cones and Pyramids. — These forms usually result from a lengthening of the lower petioles to secure a better light exposure for the under leaves, or from an increase in the size of the leaves themselves, as we see in the rosettes that form about the roots of our common biennial and perennial herbs in winter. To the same cause is due the pyramidal shape assumed by plants like the mullein and burdock, with large, undivided leaves which the light cannot strike through. The foliage on the upright stalk that rises from these rosettes in spring constantly diminishes from the ground upward, giving the plant the general outline of a sort of vegetable Eiffel tower. The upper leaves, too, will generally be found to assume a more or less vertical position so as not to cut off too much light from those below.

91. — Leaf pyramid of mullein.

92. — A plant that has been growing near an open window, showing the leaves all turned toward the light.

57. Heliotropism. — If there is any doubt about the object of all these careful adjustments it can be settled by placing any healthy young potted plant near a sunny window and at the end of a day or two, observing the position of the leaves. Then turn the pot round so that the leaves will face away from the light, and again, after a few days, observe any change that has taken place in their position. Try the experiment as often as you like and with any number of different plants, the result will be the same. This movement of plants in

93. — Rhubarb plant with leaves adjusted for centripetal drainage.

94. — A caladium showing centrifugal drainage.

response to the influence of light is called *heliotropism*, a word that means " turning to or with the sun."

58. Leaf Drainage. — Another important adjustment that leaves undergo is in regard to water. Notice the leaves of tulips, hyacinths, beets, turnips, and of bulbs and plants generally whose roots do not spread in a horizontal direction, and it will be found that their leaves usually assume a position more or less like that shown in Figure 93. Their edges are apt to curve inwards and they slope from base to apex at such an angle as to carry most of the water that falls upon them straight to the axis of growth, and so on down to the root. In most trees and shrubs,

95.— Leaf with tapering point that acts as a gutter in conducting off water.

on the other hand, and in plants generally with spreading roots, the leaves slope from base to tip so that the water is carried away from the axis to the circumference, where the delicate young root fibers grow that are most active in the work of absorption. In the first case the drainage is said to be *centripetal*, or towards the center of growth; in the second, it is *centrifugal*, or away from the center.

59. Leaf Cups.— The water could not well run down a long, slender leaf stalk from the blade to the stem, hence, in plants fitted for centripetal drainage the leaves are generally sessile, or the petioles are grooved or appendaged in various ways, as in the winged leaf stalks of the sweet pea and the common leaf cup (*Polymnia*), which takes its name from the cuplike expansion into which the base of the petiole is often dilated. Connate leaves may also serve the same purpose. Can you think of any other probable use for these natural water holders? Why, for

96. — Winged petiole of   97. — Water cups of Silphium
Polymnia.   perfoliatum.

instance, do housewives sometimes set the feet of their
cupboards in vessels of water?

60. Protection against Excessive Light and Heat. — With
plants growing in very hot, dry climates, or in exposed
situations, it is often necessary to guard against too rapid

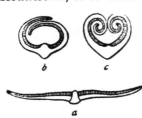

98. — Cross sections of the
leaf of sand grass : *a*, unrolled
in its ordinary position ; *b* and
c, rolled up to prevent too
rapid transpiration.

transpiration by shutting off the
direct rays of the sun from the stomata, just as we close our blinds in
summer to keep the heat out. The
common blackberry lily (*Belam
canda*) of our old red hillsides, and
others of the iris family, to which
it belongs, have their leaves ranged
vertically so as to expose only the
tips to the full glare of the noonday
sun. Many swamp herbs like the sweet flag (*Acorus
calamus*), the cat-tails, and yellow-eyed grass (*Xyris*) have
the same habit, the pools and marshes in which they grow
often becoming dry in summer ; and moreover, even though
there may be plenty of moisture, they are very dependent
upon it and need to retain a good store. Strongly revolute margins, such as are found in many sand plants
growing along the seashore, produce the same effect by
inclosing the stomata in the hollow trough or cylinder
formed by their recurved edges.

61. Compass Plants. — A very remarkable adjustment is that of the rosinweed, or- compass plant (*Silphium laciniatum*), which grows in the prairies of Alabama and westward, where it is exposed to intense sunlight. The leaves not only stand vertical, but have a tendency to turn their edges north and south so that the blades are exposed only to the gentler morning and evening rays. The prickly lettuce manifests the same habit.

99, 100. — A compass plant, rosinweed (*Silphium laciniatum*): 99, seen from the east; 100, seen from the south.

62. Leaves that go to Sleep. — The leaves of many plants change their position at night as if folding themselves for sleep. This habit is especially noticeable in certain members of the pea family and also in the wood sorrel and the cultivated oxalis of the gardens. The motions may be either spontaneous, as in the telegraph plant (*Desmodium gyrans*), or in response to various external agents, as light, heat, irritation by contact with other substances, etc. The positions assumed are various and may even differ in different parts of the same compound leaf; in

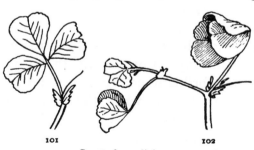

101, 102. — Spotted medick: 101, awake; 102, asleep.

the kidney bean (*Phaseolus*), for instance, the common petiole turns up at night and the separate leaflets down.

63. Experiments. — Place a healthy plant of oxalis, spotted medick, or white clover in a pot and keep it in your room for observation. Notice the changes of position the leaves undergo. Sketch one as it appears at night and in

the morning.   Can you think of any benefit a plant might
derive from this habit of going to sleep ?

In order to determine whether these changes are due to
want of light or of warmth, put your plant in a dark

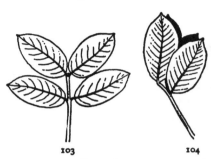

closet in the middle of the
day, without change of temperature.   After   several
hours note results.   Transfer to a refrigerator, or in
winter, place outside a window where it will be exposed
to a temperature of about
5° C. (40° F.) for several
hours, and see if any change
takes place.   Next put your

103, 104. — Ground pea or peanut:
103, in day position; 104, in night position.

plant at night in a well-lighted room and note the effect.
If practicable, keep a specimen for several months in
some place where electric lights are burning continuously
all night, and try to find out whether it is possible to kill
a plant for want of sleep.

64. Autumn Leaves. — When trees prepare for their
winter sleep the sap all retires from the foliage back to the
stem and roots, and the leaves, having no more work to do,
give up their chlorophyll and fall away.   It is this breaking
up of the chlorophyll by the oxygen of the air, that gives
to the autumn woods their brilliant coloring, and not the
action of frost.   After a wet season, when the leaves are
full of sap and nourishing juices, the chemical changes
attendant upon the withdrawal of the chlorophyll are more
active, and the changes of color more vivid than after a
period of drought, when the leaves wither and fall away
with little display of color.

65. The Physiological Significance of leaf adjustment
will be evident if we consider that the process of food
manufacture is entirely dependent upon the action of chlorophyll through the agency of light.   Without this agency
no food can be produced, though its influence is not always

direct. Seeds germinate, bulbs and rootstocks perform
their vegetative functions, and many parasites and saprophytes grow and flourish in the dark, but in these cases it
is always at the expense of reserve material provided by
the plant itself, or by the host, through the agency of
chlorophyll acting in the light.[1] It is the green leaves of
summer that lay up the stores of food in bulbs and rootstocks for winter, and
flowering stems will even
grow and blossom in the
dark if enough green
leaves are left exposed
to manufacture nourish-
ment for them.

Pass the end of a
budding flower stem of
any green-leaved plant —
gourd, squash, water
melon, morning-glory,
etc., make good examples
— through a small hole
into a dark box, leaving
the rest of the plant ex-

105. — Experiment with a gourd developed
partly in the dark and partly in the light.

posed to light, and taking care not to bruise or injure it in
any way. Cover the entire leafy portion of another plant
of the same kind with a box, leaving only the flower bud
exposed, and covering, or cutting away any new leaves
that may appear. Watch what happens, and at the end
of two or three weeks compare results. The green plant
may not show any change for several weeks, until it
has used up the chlorophyll already stored away in its
leaves.

Experiments like the foregoing show very plainly that
it is no mere figure of rhetoric to speak of the coal hidden
away in the earth as " stored up sunshine."

[1] Recent discoveries have given reason to believe that a few of the bacteria
are exceptions to this statement, but with regard to the generality of plants, it
holds true.

1. Why are the outer twigs of trees generally the most leafy? (54, 55, 56.)

2. Is the common sunflower a compass plant? Is cotton?

3. Are there any such plants in your neighborhood?

4. Compare the leaves of half a dozen shade-loving plants of your neighborhood with those of as many sun-loving ones ; which, as a general thing, are the larger and less incised?

5. Give a reason for the difference. (53, 56.)

6. Why do most leaves — notably grasses — curl their edges backwards in withering? (17, 60.)

7. What advantage is gained by doing this? (60.)

8. Observe such of the following plants as are found in your neighborhood, and report any changes of position that may take place in their leaves and the causes to which such changes should be ascribed : wood sorrel, mimosa (*Albizzia*), honey locust, wild senna (*Cassia marilandica*), partridge pea (*C. chamæchrista*), wild sensitive plant (*C. nictitans*), red bud, bush clover (*Lespedeza*), Japan clover (*L. striata*), Kentucky coffee tree, sensitive brier (*Schrankia*), ground pea or peanut, kidney bean.

9. Which of the trees named below shed their leaves from tip to base of the bough (centripetally), and which in the reverse order? Ash, beech, hazel, hornbeam, lime, willow, poplar, pear, peach, sweet gum, elm, sycamore, mulberry, China tree, sumac, chinquapin.

TRANSFORMATIONS OF LEAVES

MATERIAL. — Any kinds of leaves that can be obtained showing adaptations for protective and other purposes, such as scales, spines, tendrils, glands, etc. Some of those mentioned in the text are : sweet pea, cedar, cactus, asparagus, cabbage, stonecrop, purslane, sarracenia, bladderwort (*Utricularia*), sundew (*Drosera*), Spanish bayonet (*Yucca*), stinging nettle (*Urtica*), horse nettle (*Solanum carolinense*). The subject is best studied out of doors, or in a greenhouse.

66. Besides performing their natural functions, leaves are modified in various ways to do the work of other organs. No part of the plant is subject to more curious and varied metamorphoses, and they are made to serve all sorts of purposes.

67. Leaves as Tendrils. — Examine a leaf of the wild vetch, or of the common garden pea, and it will be seen that the two or three upper leaflets are transformed into tendrils for climbing. In the sweet pea all but the two lowest leaflets have been developed into tendrils.

68. Scale Leaves. — Sometimes the leaf disappears entirely, or is reduced to a mere scale or spine, as in the cedar and most cactuses, and some other part takes its place, but it can always be recognized by its position on the stem, just below the point where a bud appears. Ordinarily, buds never occur anywhere except at the axil, and this

106. — Leaf of common pea, showing upper leaflets reduced to tendrils.

position is so constant that it will generally serve to distinguish leaves from other organs under all disguises. In the common asparagus, the green threadlike appendages which are usually regarded as foliage, spring each from the axil of a little scale. This, as has just been stated, is the normal position of a bud or branch, and hence, botanists conclude that here a double transformation has taken place, of leaves into scales and branches into foliage.

Scale leaves are of use to plants that have need to protect themselves against frost and snow, like the heaths and mosses of cold regions. They are common also in hot and arid districts where it is necessary to reduce the surface exposed for transpiration, though here they are more apt to take the form of prickles and spines as a double protection against sun and animals.

69. Leaves as Storehouses of Food and Moisture. — Of this we have familiar examples in the cabbage and other salad plants of the garden. In some of the fleshy stone-crops and purslanes, the leaves seem to have transformed themselves into living water bags.

70. Death Traps. — The sarracenia, better known as the pitcher plant, or trumpet leaf, is a familiar example of these vegetable insect catchers. Its curious pitcher-shaped, or trumpet-shaped leaves are traps for the capture of the small game upon which the plant feeds. The lower part of the blade is transformed into a hollow vessel for holding water, and the top is rounded into a broad flap called the *lamina*. Sometimes the lamina stands erect, as in the common yellow trumpets of our coast regions, and when this is the case, it is brilliantly colored and attracts insects. Sometimes, as in the parrot-beaked and the spotted trumpet leaf (Fig. 108), it is bent over the top of the water vessel like a lid, and the back of the leaf, near the foot of the lamina, is dotted with transparent specks that serve to decoy foolish flies away from the true opening and tempt them to wear themselves out in futile efforts to escape, as we often see them do against a window pane.

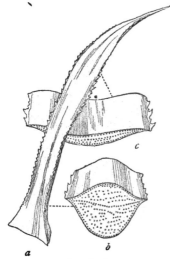

107. — *a*, Leaf of an agave, or American aloe, thickened for the storage of water; *b* and *c*, cross sections made at points indicated by the dotted lines.

108. — Spotted sarracenia (*S. vari-olaris*): *l*, lamina; *s*, transparent spots (*after* GRAY).

If the contents of one of these leaves are examined with a lens there will generally be found mixed with the water at the bottom, the remains of the bodies of a large number of insects. Notice that the hairs on the outside all point up, towards the rim of the pitcher, while those on the inside turn downward, thus smoothing the way to destruction but making return impossible to a small insect when once it is ensnared. When we remember that these plants are generally found in poor, barren soil, we can appreciate the value to them of the animal diet thus obtained.

71. Other Examples of insect-catching leaves are the Venus's flytrap, found nowhere but in a certain section of North Carolina, near the coast, and the little sundew (*Drosera rotundifolia*), which Mr. Darwin has made the heroine of his famous book on "Insectivorous Plants." It is a delicate, innocent looking little flower, and owes its poetic name to the dewlike appearance of a shining, sticky

109. — Plant of sundew.

fluid exuded from the glands on its leaves, which glitter in the sun like diamond dewdrops. It is, however, the most

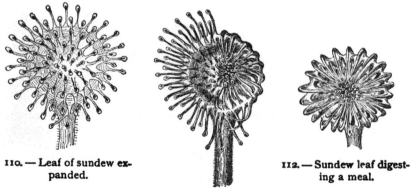

110. — Leaf of sundew expanded.

112. — Sundew leaf digesting a meal.

111. — Leaf closing over captured insect.

110–112. — Leaves of sundew magnified.

voracious of all carnivorous plants, the shining, sticky leaves acting as so many bits of fly paper by means of which it catches its prey. When a fly has been trapped,

the edges of the leaf curve inwards, making a little pouch
or stomach, and an acid juice exudes from the glands and
digests the meal.   After a number of days, varying
according to the digestibility of the diet, the blades slowly
unfold again and are ready for another capture.

In the bladderwort, common in pools and still waters
nearly everywhere, the petioles are transformed into floats,
while the finely dissected, rootlike blades bear little bladders
which, when examined under the microscope, are found to
contain the decomposed remains of captured animalculæ.

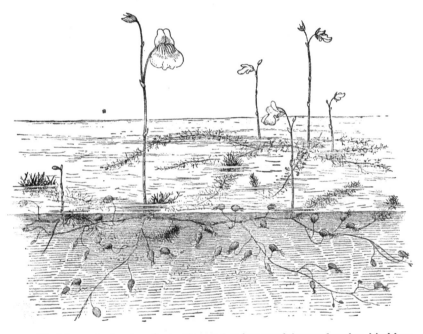

113.— Bladderwort, showing finely dissected submerged leaves bearing bladders,
and petioles transformed to a whorl of floats for buoying up the flowering stem.

72. Protective Leaves. — One of the most frequent
modifications of leaves is for protection, either of themselves or of other organs, against animals, drought, excessive moisture, dust, heat, cold, etc.   The prickles of the
thistle and horse nettle, the hairs of the stinging nettle,
and the sharp spears from which the Spanish bayonet
(*Yucca aloifolia*) takes its name, are all familiar examples
of the first kind, as are also the venom of the poison ivy,

the fetid odors of the jimson weed and China tree, and even the aroma of the pennyroyal and lavender that we pack with our clothes in summer to keep the moths away.

114. — Protective awl-pointed leaves of Russian thistle.

115. — Spine protected leaf of horse nettle.

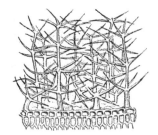

117. *Verbascum Thapsiforme.*

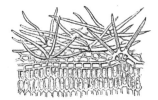

118. *Potentilla cinerea.*

119. *Shepherdia.*

116. — Spearlike leaves of Spanish bayonet. 117–119. — Protective hairs magnified.

The protective devices of leaves are generally so apparent that the student can easily make them out for himself, with the help of a few suggestive questions.

PRACTICAL QUESTIONS

1. How can it benefit a plant to have its leaves, or some of them, changed to tendrils? (67.)

2. What advantage to plants is it to be able to climb? (54–57, 65.)

3. Why is it that evergreen trees and shrubs have generally either thick, hard, coriaceous leaves, like those of the holly and magnolia, or scales and needles, as in the cedar and pine? (68.)

4. Why are winter herbs with tender foliage, like the chickweed and winter cress, generally low stemmed, and disposed to keep close to the earth?

5. Why do many plants which are *deciduous* — that is, shed their leaves in winter — at the north, tend to become evergreen at the south?

6. Question 5 seems to conflict with question 13, page 27; can you reconcile them?

7. Can you find any kind of leaf that is not preyed upon by something? If so, how do you account for its immunity?

8. Make a list of some of the most striking of the protected leaves of your neighborhood.

9. What is the nature of the protective organ in each case?

10. For protection against what does it seem to be specially adapted?

11. Are the plants in your list for the most part useful ones, or troublesome weeds?

12. Examine the leaves of the worst weeds that you know and see if these will help in any way to account for their persistency.

FIELD WORK

The study of this subject and of all those that follow should be supplemented by field work, in expeditions organized for the purpose; furthermore, the student can learn a great deal for himself by keeping his eyes open and observing the plants he meets with in his ordinary walks.

In connection with Sections 14–30, consider the effects upon soil moisture of water transpiration from the leaves of forest trees that strike their roots deep, and from those of shallow-rooted herbs and weeds that draw their water supply from the surface. Consider the value of forests in protecting crops from excessive evaporation by acting as wind breaks. Study the effect of the fall of leaves on the formation of soil.

In any undisturbed forest tract turn up a few inches of soil with a garden trowel and see what it is composed of. Notice what kind of plants grow in it. Note the absence of weeds and account for it. Compare the appearance of trees scattered along windy hillsides, where the fallen leaves are constantly blown away, or in any position where the soil is unrenewed, with those in an undisturbed forest, and then give an opinion as to the wisdom of hauling away the leaves every year from a timber lot.

Sections 31–49. Observe the effect of the lobing and branching of leaves in letting the sunlight through. Notice any general differences that may appear as to shape, margin, and texture in the leaves of sun plants, shade plants, and water plants, and account for them. Study the arrangement of leaves on stems of various kinds and see how it is adapted in each case to the shape of the foliage. Consider the value of the various kinds of foliage for shade; for ornament; as producers of moisture; as food; as insect destroyers, etc.

It is important to learn to know and distinguish the different kinds of trees and shrubs in your neighborhood by their leaves. A useful exercise for this purpose is to make a collection of those of some family like the oaks or hawthorns, that contains a great many varieties, and compare them carefully with one another.

Sections 50–65. In different mosaics and rosettes of leaves study the means by which the adjustment has been brought about and the purpose it subserves. Notice the form and position of petioles of different leaves, and their effect upon light exposure, drainage, etc., and the behavior of the different kinds in the wind. Look for compass plants in your neighborhood, and for other examples of adjustment to heat and light. Study the position of leaves at different times of day and in different kinds of weather and note what changes occur and to what they are due. The sunflower and pea families offer some of the most striking examples of this kind of sensitiveness. The oxalis and geraniums, cotton, and others of the mallow family ought also to be investigated.

Study the drainage system of different plants and observe whether there is any general correspondence between the leaf drainage and the root systems. This will lead to interesting questions in regard to irrigation and manuring. (Where plants are crowded the growth of both roots and leaves is complicated with so many other factors that it is best to select for observations of this sort specimens growing in more or less isolated situations.)

Notice the time of the expansion and shedding of the leaves of different plants, and whether the early leafers, as a general thing, shed early or late; in other words, whether there seems to be any general time relation between the two acts of leaf expansion and leaf fall.

Sections 66–72. Look for instances of protected leaves; study the nature and position of the protective organs and decide as to their special purpose, whether as defenses against heat, cold, dust, or insects and other animals.   Examine the tendrils of various climbing plants and tell from their position whether they represent stipules, leaves, or branches. Look for instances of transformed leaves of any kind and try to understand the nature and object of the transformation.   Always be on the alert for transformations and disguises everywhere, and account for them as far as you can.

III. FRUITS[1]

FLESHY FRUITS

MATERIAL. — Apple, pear, haw, hip, or other pome fruit; any kind of melon or gourd fruit (if a specimen of the turban squash can be obtained it will illustrate well the morphology of this kind of fruit); tomato, cranberry, lemon, grape, or other kind of berry; a pickled peach or cherry, or some kind of wild drupe, as dogwood or black haw. City schools can obtain specimens for the lessons in this chapter from fruit stores, and teachers can do a great deal by collecting and preserving material when on their summer outings.

73. What is a Fruit? — The word fruit does not mean exactly the same thing to the botanist that it does to the gardener and the farmer. Botanically, a fruit is any ripened seed vessel, or *ovary*, as it is technically named, with such connected parts as may have become incorporated with it; and so, to the botanist, a boll of cotton, a tickseed, or a cockle bur is just as much a fruit as a peach or a watermelon.

74. The Pome. — Examine an apple or pear. With the point of a pencil separate the little dry, pointed scales that cover the depression in the center of the end opposite to the stem. These are the remains of the *sepals*, or lobes of the little green cup called a *calyx* that will be found at the base of all apple and pear blossoms in spring. Their

[1] It may seem a little premature to begin the study of fruits here, as some kinds cannot be fully understood without examining them in connection with the flower, but the desirableness of taking them up at a season when material is abundant seems to the author more than an offset to this objection. It will be found a great advantage, moreover, to familiarize the pupil with the structure of the ripened ovary, where the parts are large and easy to distinguish, before taking up the study of that organ, in the flower, where it is often so small that it can not conveniently be dissected.

nature will be more apparent on comparing them with
a hip, which is clearly only the end of the footstalk
enlarged and hollowed out with
the calyx sepals at the top. Cut
a cross section midway between
the stem and the blossom ends,
and sketch it. Label the thin,
papery walls that inclose the seed,
carpels. How many of them are
there, and how many seeds does
each contain? The carpels taken
together constitute the *pericarp*,
or wall of the seed vessel. The

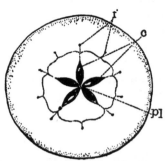

120.—Cross section of a pome:
pl, placenta; *c*, carpels; *f*, fibrovascular bundles.

fleshy part of the apple is, strictly speaking, no part of
the seed vessel or ovary proper, but consists merely of the
receptacle, or end of the footstalk, which becomes greatly
enlarged and thickened in fruit. The word pericarp, however, is often taken in a broader sense, to include all that
portion of the fruit which surrounds and adheres to the
ovary, no matter what its nature or texture. Look for a
ring of dots outside the carpels, connected (usually) by
a faint scalloped line. How many of these dots are there?
How do they compare in number with the carpels? With
the remnants of the sepals ad-
hering to the blossom end of
the fruit?

75. Next make a vertical section through a fruit, and sketch
it. Notice the line of woody
fibers outside the carpels, inclosing the core of the apple. Compare this with your cross section;
to what does it correspond?
Where do these threads originate? Where do they end? Can

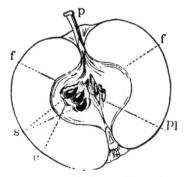

121.—Vertical section of a
pome: *p*, peduncle; *f*, fibrovascular bundles; *s*, seeds; *pl*, placenta; *c*, carpel.

you make out what they are? (See Section 43; they are the
fibrovascular bundles that connected the veins in the petals

and sepals of the apple blossom with the stem.) Notice how and where the stem is attached to the fruit. Label the external portion of the stem *peduncle*, the upper part, from which the fibrovascular bundles branch, the *torus*, or *receptacle*. It is the enlargement of this which forms the fleshy part of the fruit.[1] Try to find out, with the aid of your lens and dissecting pins, the exact spot at which the seeds are attached to the carpels, and label this point *placenta*. Notice whether it is in the axis where the carpels all meet at their inner edges, or on the outer side. Observe, also, whether the seed is attached to the placenta by its big or its little end. If you can find a tiny thread that attaches the seed to the carpel, label it *funiculus*, or seed stalk.

76. Use of the Rind. — Select two apples of equal size, peel one, and then weigh both. After twelve to twenty-four hours, weigh them again. Which has lost most? What is the use of the rind? Place peeled and unpeeled fruits in an exposed place and see which is most readily attacked by insects. Which decays soonest?

Write under the sketches that you have made the word *pome*, which is the botanical name for this kind of fruit. Write a definition of a pome.

77. Modifications of the Pome. — Compare with the drawings you have made, a haw and a hip. What points of agreement do you see? What differences? Which of the two more closely resembles the typical pome?

The pome is not the only fruit of which the receptacle forms a part. Other well-known instances of this sort of modification are the fig, lotus, and calycanthus (see Figs. 123, 124); but a fruit is not a pome unless the containing receptacle becomes more or less soft and edible. The receptacle is subject to a great variety of modifications and forms a part of many fruits.

---

[1] See *Pome*, "American Encyclopedia of Horticulture," Macmillan Co.

122. — Vertical section of a hip, showing seeds contained in a hollow receptacle (*after* GRAY).

123, 124. — Enlarged receptacle of Carolina allspice (*Calycanthus*) containing fruits attached to its inner surface: 123, exterior; 124, vertical section.

78. The Pepo, or Melon. — Next examine a gourd, cucumber, squash, or any kind of melon, and compare its blossom end with that of the pome. Do you find any remains of a calyx, or other part of the flower? Examine the peduncle and observe how the fruit is attached to it. Cut cross and vertical sections, and sketch them, labeling each part. There may be some difficulty in making out the carpels, for they are not separate and distinct as in the pome, but confluent with the enlarged receptacle, which in these fruits forms the outer portion of the rind,[1] and also with each other at their edges, so as to form one unbroken circle, as if they had all grown together. And this is precisely what has happened. The number of carpels can easily be distinguished, however, by counting the placentas, which divide the interior into compartments called *cells* or *loculi*, corresponding in number to the number of carpels. The placentas are greatly enlarged and modified, and it may be necessary to refer to the diagram, Figure 125, in order to make them out. How many cells, or chambers,

125. — Cross section of gourd: *c*, one of the carpels in diagram (*after* GRAY).

are there in your specimen? How many placentas? Are the seeds vertical, as in the apple, or horizontal? Look

[1] See *Cucurbita*, "American Encyclopedia of Horticulture."

for the little stalk, or thread, that attaches them to the placenta.

Pepo is the name given by botanists to this kind of fruit. Write in your notebook a proper definition of it, from the specimens examined.

79. The Berry. — Examine a tomato, an eggplant, a grape, cranberry, lemon, or orange, in both cross and vertical section, and compare it with the pepo. Notice that they all agree in having a more or less thick and firm outside covering filled with

126   127

126, 127. — A potato berry: 126, exterior; 127, cross section.

a soft, pulpy interior. In what respects does the one you are examining differ from the pepo?

Fruits of this kind are classed by botanists as berries. They are the commonest of all fleshy fruits, and the most variable and difficult to define. In general, any soft, pulpy, or juicy mass, like the grape and tomato, whether one or many seeded, inclosed in a containing envelope, whether skin or rind, is a berry. Its typical forms are such fruits as the grape,

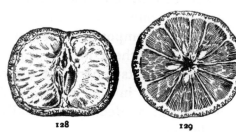

128   129

128, 129. — Tangerine: 128, vertical section; 129, cross section.

mistletoe, pokeberry, etc., though such diverse forms as the eggplant, persimmon, red pepper, orange, banana, and pomegranate have been classed as berries.; and, in fact, the pepo itself is only a greatly modified kind of the same fruit. In popular language, any small, round, edible fruit is called a berry, but do not confound it with

80. The Drupe, or stone fruit, of which the cherry, plum, peach, dogwood, black haw, and black gum furnish typical examples.

Notice that the drupe agrees with the berry in having

a more or less juicy or fleshy interior surrounded by a protecting skin, but the stone within this is not a mere seed, such as we find in the berry, but consists of the inner layer of the pericarp, which has become hard and bony. Open the stone and the seed will be seen with its own coverings inside. Have you ever found a stone with more than one kernel to it; for instance, in eating almonds? This fact shows that the stone is not a seed coat, but the hardened inner wall of a seed vessel or ovary; for a seed coat can never contain more than one seed any more than the same skin can contain more than one animal. In a green drupe, before the stone has hardened, its connection with the fleshy part is very evident. This stony layer enveloping the seed is the main distinction between the drupe and the berry, and it is not always . possible to make it out except by an examination of the young ovary. Of course there can be but one stone to a carpel, as each carpel has only one inner coat to be hardened; but where a drupe is composed of several carpels clustered together, as we saw them in the apple, each one may produce a stone from its inner coat while the outer coats become confluent, as in the melon, and in this way a drupe may be several seeded, as is actually the case in the dogwood, elder, etc.

130. — Vertical section of a drupe (*after* GRAY).

All the fruits that have been considered in Sections 73–80 belong to the class of fleshy ones. These form the great bulk of the fruits sold in the market and served upon our tables, and are of special importance to the horticulturist.

PRACTICAL QUESTIONS

1. Examine such of the fruits named below as you can obtain, and tell to which of the four kinds described each belongs: asparagus, horse nettle, China berry, smilax, hackberry, pawpaw, guava, persimmon, red pepper, orange, buckeye, gherkin, pumpkin, prickly pear, mangrove, whortleberry, banana, date, olive, maypop, cedar berry, Ogeechee lime.

2. Which are the commonest of fleshy fruits in autumn?

3. Name six of the most watery fruits that grow in your neighborhood.

4. Under what conditions as to soil, heat, moisture, etc., does each thrive best?

5. Would a gardener act wisely to infer that because a fruit contains a great deal of water it should be planted in a very wet place?

6. Which contains most water, the fruit or the leaves of the apple?

7. Why does the fruit not wither when separated from the tree, as the leaves do? (76.)

DRY FRUITS

MATERIAL. — Acorn or other nut; a cotton boll or a pea or bean pod; various small, seedlike fruits, such as the so-called seeds of the sunflower, carrot, parsley, clematis, grains of corn, etc.

81. Importance of Dry Fruits. — Dry fruits are not in general so conspicuous or attractive as fleshy ones, but on account of their greater number and variety they offer a wide field for study. And when we consider that the grains which furnish our breadstuffs, and the beans and nuts that form so large a part of our food all belong to this class we realize that they have an even greater claim upon our attention than the most brilliant products of the garden.

82. Different Kinds of Dry Fruits. — Compare an acorn, a chestnut, or a hazelnut with a ripe cotton boll or a bean pod. Try to open each with your fingers; what difference do you perceive?

This difference gives rise to the distinction of dry fruits into

83. Dehiscent: those that open at maturity in a regular way for the discharge of their seed; and

84. Indehiscent: those that remain closed until the dry carpels are worn away by decay, or burst by the germination of the contained seed.

85. Why Some Fruits Dehisce. — Open each of your specimens; how many seeds, or kernels, does the indehiscent one contain? The dehiscent one? Can you explain

now why the one should open and the other not? Would
it be of any advantage for a one-seeded pod to open?
Remove the kernel from the indehiscent fruit; has it any
covering besides the shell? Which is the pericarp, and
which the seed coat?

86. Indehiscent Fruits are so simple that it will not be
necessary to devote much time to them. Gather specimens
of as many kinds as you can find, and try to identify them
by means of the pictures and descriptions that follow. Do
not try to memorize these descriptions, but use them merely
as a help in studying actual specimens. The acorn, hickory nut, chestnut, etc., furnish good examples of

87. The Nut, which is easily recognized by its hard,
bony covering, containing usually, when mature, a single
large seed that fills the interior. Care must be taken not
to confound with true nuts, large bony seeds, like those of
the buckeye, horse-chestnut, date, and the Brazil nut sold

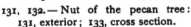

131, 132.—Nut of the pecan tree: 133, 134. — Nutlike seeds: 133,
131, exterior; 133, cross section. horse-chestnut; 134, seed of *sterculia
fœtida.*

in the markets. In the true nut the hard covering is the
seed vessel, or pericarp, and no part of the seed itself,
though it often adheres to it so closely as to seem so. In
bony seeds like those of the horse-chestnut and persimmon
the hard covering is the seed coat. The distinction is not
always easy to make out unless the seed can be examined
while still attached to the placenta of the fruit.

88. The Achene, of which we have examples in the tailed
fruit of the clematis, the tiny pits on the strawberry, and

the so-called seeds of the thistle, dandelion, etc., is a small,
dry, one-seeded indehiscent fruit,
so like a naked seed that it is
generally taken for one by persons who are not acquainted with
botany. It is the commonest of
all fruits, and there are so many
kinds that special names have been
applied to some of the most marked
varieties. The achene of the composite family may generally be

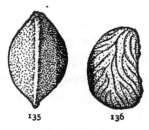

135, 136. — Achenes (magnified): 135, of buckwheat; 136,
of cinquefoil.

known by the various appendages
in the form of scales, hooks, hairs,
or chaff, that crown it (Figs. 137–
142). This appendage is always
called a *pappus*, no matter under
what form it occurs. It is fre-

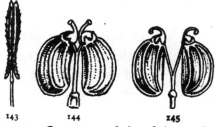

137–142. — Achenes of the composite family (GRAY): 137, mayweed (no pappus); 138, chicory (its pappus a shallow cup); 139, sunflower (pappus of two
deciduous scales); 140, sneezeweed (*Helenium*), with its pappus of five scales;
141, sow thistle, with its pappus of delicate downy hairs; 142, dandelion, tapering
below the pappus into a long beak.

quently deciduous, as in the sunflower, and sometimes
wanting altogether, as in
the mayweed.

143–145. — Cremocarps, fruits of the parsley
family.

89. Cremocarp is the
name given to the fruit
of the parsley family.
It is merely a sort of
double achene attached
by the inner faces to a
slender stalk called the *carpophore*, or carpel bearer, from

which it separates at maturity. Gather a fruiting cluster of fennel, parsley, caraway, etc., and examine one of the small seedlike fruits through a lens. Separate the two achenes of which it is composed, and find the carpophore between them. Sometimes it splits in two (Fig. 145), one half going with each achene; or they may separate from it through their entire length and remain suspended from the top (Fig. 144). Notice the longitudinal ribs on the back of the achenes, or *mericarps*, as they are called. Between these ridges are situated the *vittæ*, or oil tubes to which the aromatic flavor of these fruits is due.

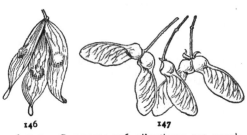

146   147
146, 147. — Samaras: 146, ailanthus; 147, maple.

90. The Samara, or key fruit, is an achene provided with a wing to aid in its dispersion by the wind. The maple, ash, elm, etc., furnish familiar examples.

91. The Grain, or caryopsis, so familiar to us in all kinds of grasses, is a modification of the achene in which the seed coats have so completely fused with the pericarp that they can no longer be distinguished as separate organs. Peel the husk from a grain of corn that

148   149
148, 149. — Grain of broom corn millet: 148, front view; 149, back view.

has been soaked for twenty-four hours, and you will find the contents exposed without any covering; remove the shell of an acorn or a hickory nut, and the seed will still be enveloped by its own coats. Would it be any advantage for the seed of an indehiscent fruit, like a grain of corn or oats, to have a special covering of its own?

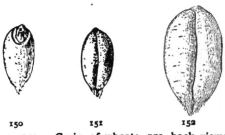

150   151   152
150–152. — Grain of wheat: 150, back view; 151, front view; 152, front view (magnified).

92. Distinction between Nuts and Achenes. — In very small fruits it is not easy to distinguish between a nut and an achene, nor is it very material. Technically, an achene is a fruit composed of a single carpel, a nut of two or more which have become so completely fused together that their separate parts can be detected only by examining the unripe seed vessel in the flower. Botanists apply the terms very loosely, and the beginner need not be distressed if he can not classify exactly all the specimens he meets with. In general, the larger, harder, and bonier fruits of the kind are called nuts. The family to which a specimen belongs must also be taken into consideration. For instance, the achene being the characteristic fruit of the sunflower family, any puzzling specimen of that family, like the cockle bur, would naturally be classed as an achene.

PRACTICAL QUESTIONS

1. Name all the indehiscent fruits you can think of that are good for food or other purposes.

2. Make a list of the commonest indehiscent fruits of your neighborhood.

3. Which of these are useful for any purpose?

4. Which are troublesome weeds?

DEHISCENT FRUITS

MATERIAL. — Simple follicles of larkspur, milkweed, etc.; a pod of pea or bean; pods of any species of the mustard family, or of the trumpet vine (*Tecoma*); cotton, okra, iris, or Indian shot (*Canna*). Cotton or okra are preferable if they can be obtained, because the parts are large and well defined.

93. Simple, or Monocarpellary Fruits. — Pod, or capsule, is the general name given to all dehiscent fruits. The latter term is properly confined to pods of more than one carpel, but the distinction is not strictly observed by botanists. The simplest possible kind of a pod is

94. The Follicle, of which the larkspur, milkweed, marsh marigold, etc., are familiar examples. It is composed of a single carpel, which may be regarded as a modified leaf. Examine one of these pods and you will find that it splits down one side, which corresponds to the edges of the leaf brought together and turned inwards to form a placenta for the attachment of the seed. This line of union is called a *suture*, from a Latin word meaning a seam.

153. — Follicle of milk-weed.

95. The Carpel a Transformed Leaf. — The leaflike nature of the carpel is very evident in such fruits as the follicles of the Japan varnish tree (*Sterculia platanifolia*), where even the veining is quite distinct, and the whole carpel so leaflike in appearance that there is no mistaking its nature. Indeed, after the wonderful transformations we have already found leaves undergoing, their development into the hardest and thickest of carpels need not surprise us.

154. — Leaflike follicle of Japan varnish tree: *s, s*, sutures.

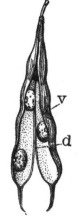

155. — Legume of bean: *v*, ventral suture; *d*, dorsal suture.

96. The Legume. — Get a pod of any kind of bean or pea, and observe that it differs from the follicle in having two sutures or lines of dehiscence. One of these, which runs along the back of the carpel and corresponds to the midrib of the leaf, is called the outer, or *dorsal*, suture; the other, corresponding to the united edges of the carpellary leaf, is the inner, or *ventral*, suture, so called because it always turns inwards, that is, towards the center or axis of the flower.

97. Origin of the Name.—This kind of pod is the characteristic fruit of the great pea, or pulse family, and gets its name from the Latin word, *lego*, to pick, or gather, because crops of pulse have always been picked by hand instead of being cut or mown like grain and hay.

98. Sutures.—Place a legume upon one side and sketch it, labeling the sutures. If you cannot tell which is the dorsal and which the ventral, open the pod and observe where the seeds are attached; this is the ventral suture, because in all normal carpels it is the united edges of the leaf margins, or in other words, the ventral suture, that forms the seed-bearing surface, or placenta.

99. Valves.—Sketch the open pod with the seeds in it, showing their point of attachment. Label this the *placenta*, and the two halves into which the pod has split, *valves*. Notice that the valves are not separate carpels, but only two halves of the same carpel. What is the difference between a legume and an ordinary follicle?

157.—Legume of a pea, with partially constricted pod.

158.—Loment of beggar's ticks.

156.—Constricted legume of cassia nelsoni.

100. The Loment, so unpleasantly familiar to most of us in the beggar's ticks tribe, is merely a kind of legume constricted between the seeds and breaking up into separate joints at maturity. What kind of indehiscent fruits do the joints resemble when separated?

101. The Silique is the characteristic fruit of the mustard family, as the legume is of the pea tribe, though it is

common in other plants also, the trumpet flower (*Tecoma
radicans*) being a conspicuous example. Open any convenient specimen and notice the manner of dehiscence.
How does it differ from that of the legume? What other
difference do you perceive? Are the edges of the valves
reflexed or folded in any way so as to form the two cells
or chambers into which the silique is divided? How is the
partition made? A dividing wall of this sort, that is made
in any other way than by the inflexed margins of the carpels, is called a *false partition*. Sketch
your specimen as it appears with one
of the valves removed, showing the
position and attachment of the seeds.
Where is the placenta? Is the false
partition parallel with the valves or at
right angles to them? Compare it in
this respect with other specimens of the
same family, and with the silique of the
trumpet vine, if you
can get one; is the
direction of the partition always the same?
Does it fall away with
the valves or remain
attached to the receptacle?

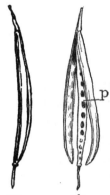

159   160
159, 160.— Silique of
mustard: 159, closed;
160, after dehiscence,
showing false partition, *p*.

161   162
161, 162.— Silicle of
shepherd's purse: 161,
entire; 162, with one
carpel removed, showing attachment of
seeds, and the false
partition running contrary to the flattened
sides.

102. **The Silicle** is only a short and
broad silique, like those of the shepherd's purse (*Capsella bursa-pastoris*)
and pepper grass (*Lepidium*). The last
two named belong to the class known as

103. **Syncarpous or Compound Pods.** — Generally speaking, there are never more carpels in a pod than there are
seed-bearing sutures. In a boll of cotton, or a pod of okra,
iris, or other large dehiscent fruit, notice the lines or seams
running from base to apex of the pericarp; into how many
sections or carpels do they divide it? When several carpels unite in this way into one body, they form a *syncar-*

pous pod or capsule—the word "syncarpous" meaning "of united carpels." The three large, leaflike bodies at the base of the cotton boll (none in the okra — unless very immature pods are used — or the iris) are *bracts*, and together they form an *involucre*. Remove these and also the remains of the flower cup, or calyx, that will be found just within them, and notice the round, flattish expansion of the stem where the fruit is attached. Make a sketch of the closed capsule, labeling this expansion *receptacle*, the stem itself *peduncle*, the longitudinal lines *sutures*, and the spaces between them *carpels*.

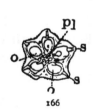

163–166. — Capsule of okra: 163, entire, *c, c*, carpels, *r*, receptacle, *s, s*, sutures; 164, vertical section, *pl*, placenta, *o, o*, ovules, *f, f, faniculus*, or seed stalk; 165, single carpel; 166, cross section, *pl*, placenta, *o, o*, ovules, *s, s*, sutures.

Open the boll, or take one that has already dehisced, remove the lint with the seed from two of the carpels, allowing them to remain in the others, and sketch the whole as it appears on the inside. Notice the protruding ridge down the center of each carpel which divides the fruit, when closed, into separate chambers or cells. Find out to what part the seeds are attached and label it *placenta*. The little threadlike stalks that attach the seed are very small and hard to distinguish from the fleece, but when they are broken away, their place can generally be detected by small, toothlike projections on the placenta.

In pods like those of okra, cotton, iris, etc., the placenta is said to be *parietal*, from a Latin word meaning a wall, because it projects from the wall of the seed vessel. From which suture does it arise, the dorsal or the ventral? Which kind of sutures are those shown on the exterior of the boll? (Secs. 94, 98). Does it dehisce by the dorsal or the ventral sutures? Notice that when a capsule splits along its dorsal sutures in this way, the segments into which it divides are made up of the two contiguous halves of adjacent carpels, just as if we should fasten a number of leaves together by their edges and then split them down their midribs, we should get an equal number of sections made up of the adjacent halves of different leaves. And on the supposition that carpels are altered leaves, this is precisely what happens in the case of syncarpous capsules such as we have been examining.

104. Modes of Dehiscence. — Make a diagram of the mode of dehiscence of your specimen, and compare it with that of a pod or the castor bean, jimson weed, St. John's-wort, flax, etc. ; or if specimens cannot be obtained, with the accompanying diagrams. What difference do you perceive in their modes of dehiscence? The first of these is called

105. Loculicidal (Fig. 167), because it splits through the back of the carpels directly into the cells or *loculi*, a word meaning "little chambers." The second is

167-170. — Diagrams of dehiscence (*after* GRAY): 167, loculicidal; 168, septicidal; 169 and 170, septifragal.

106. Septicidal, that is, the dehiscence takes place through the *septa*, or partitions that divide the cells (Fig. 168). Either of these modes may become

107. Septifragal, as in the morning-glory, where the carpels break away from the division walls, leaving them attached to the axis of the fruit (Figs. 169 and 170). Another common form is the

108. Circumscissile, in which the upper part of the pod comes off like the lid of a dish, as in the purslane, plantain, henbane, amaranth, etc.

171, 172. — Circumscissile capsule of anagallis : 171, closed ; 172, open.

109. Union of Carpels. — The carpellary leaves may unite either by their open edges, as if a whorl like that represented in Figure 77, were to grow together by the margins (Fig. 173); or each may first roll itself into a simple follicle like the larkspur and columbine (Fig. 175), and then a number of these may unite by their ventral sutures into a single syncarpous

173. — Plan of one-celled ovary formed by the union of open carpellary leaves (GRAY).

174. — Cross section of one-celled syncarpous capsule of frostweed, with parietal placentæ (GRAY).

175. — Follicles of larkspur borne on the same torus, but distinct.

capsule, with as many cells as there are carpels (Fig. 177). The seed-bearing sutures being all brought together in the center, the placenta becomes *central* or *axial*. In the first

176. — Pods of echeveria, contiguous, but distinct.

177. — Capsule of colchicum, with carpels united into a syncarpous pod.

178. — Capsule of corn cockle, with free axile placenta.

case (Fig. 174) the open carpels form a one-celled capsule, though the placentas sometimes project, as in the cotton and okra, so far as to produce the effect of true partitions with central placenta (Fig. 164). In one-celled capsules, the number of carpels can generally be determined by the number of sutures or of placentas. The placenta is not always formed by the margins of the carpels, however, but sometimes the seeds are borne upon a prolongation of the receptacle, as in the pink and the corn cockle (Fig. 178), forming a *free central* placenta. A free central placenta may also be formed when the carpels of a pod break away from the seed-bearing surface, as in Figures 169 and 170.

PRACTICAL QUESTIONS

1. Can you name any syncarpous, or compound capsule that is single-seeded?

2. Can you name any indehiscent fruit that has more than one seed?

3. Name the weeds of your neighborhood that are most troublesome on account of their adhesive fruits.

4. Do they belong, as a general thing, to the dehiscent or the indehiscent class?

5. Give a reason for these facts.

ACCESSORY, AGGREGATE, AND COLLECTIVE FRUITS

MATERIAL. — If snake strawberries (*Fragaria indica*), Osage orange, and other late fruits of the kind cannot be obtained, a pineapple may be used for the whole class. Fresh figs, if they can be obtained, make good objects for study, but dried ones may be used. Hips, haws, etc., are always plentiful at this season. As many of the fruits mentioned in the *Practical Questions* as can be obtained should be studied either in or out of class.

110. Besides the varieties already named, all fruits, whether fleshy or dry, may be either simple, accessory, aggregate, or collective. The first kind need no explanation; they consist merely of a single ripened ovary, whether of one or more carpels, as the peach, cherry, bean, lemon, etc.

111. **Accessory Fruits** are so called because some other part than the seed vessel, or ovary proper, is coherent with or accessory to it in forming the fruit, as we saw in

the apple and the hip. The accessory part may consist of any organ, but is more frequently the calyx or the receptacle. In the strawberry, the little hard bodies, usually called seeds, that dot the surface, are the true fruits (achenes). A vertical section through the center will show the edible part to consist wholly of the enlarged receptacle. In the pineapple, the edible stalk may be traced straight through a mass of

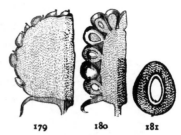

179   180   181

179–181. — Sections of accessory fruits of strawberry and blackberry, showing enlarged receptacle (*after* GRAY): 179, strawberry; 180, blackberry; 181, separate drupe of blackberry (magnified).

flowers whose seed vessels have become enlarged and ripened into fruits. Some accessory fruits, the strawberry and blackberry for example, are, at the same time,

182   183   184

182–184. — Aggregate fruit of magnolia umbrella (*after* GRAY): 182, ripened cone with a seed hanging from a lower dehiscent carpel; 183, vertical section; 184, separate follicle.

112. Aggregate, that is, they are composed of a number of separate individual fruits produced from a single flower. The cone of the magnolia and of the wild cucumber are aggregate fruits; can you name any others? The pineapple, on the other hand, is both an accessory and a

113. Collective, or Multiple Fruit, being composed of the ripened seed vessels or ovaries of a number of separate flowers that have become more or less coherent. The osage orange, sweet-gum balls, fig, and mulberry are of this class.

114. Flower Fruits. — Compare a section through a fig with those of the hip and calycanthus (Figs. 122, 124). Of what is the part that we call the skin a modification? Observe that here the receptacle is modified to a greater degree than in the rose and calycanthus, forming a sort of closed urn in which the flowers are contained. Examine

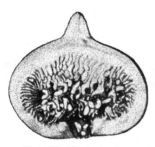

185. — Vertical section of a fig, showing the minute flowers inside the closed receptacle.

the contents with a lens, and it will be found that they consist of hundreds of what appear to be tiny seeds enveloped in a pulpy mass, but which are, in reality, the small achenes produced by a multitude of minute flowers that line the receptacle. In the fig this entire mass becomes pulpy and edible at maturity, so that we only state a fact when we commit the hibernicism of saying that the fruit of the fig is a flower — or rather, a bunch of flowers. The same is true of the mulberry, only here the edible flower mass is attached not to the inside, but to the outside of the receptacle.

The fig and strawberry are both accessory fruits; which of them is collective also, and which aggregate? The mulberry and blackberry? Is it possible always to distinguish between an aggregate and a collective fruit without having examined the flower?

115. Fruit Clusters. — Be careful not to confound aggregate and collective fruits with mere clusters like a bunch of grapes or of sumac berries. The distinction is not always easy to make out. The clump of achenes that make up a dandelion ball, for instance, though held on a common receptacle, like the mulberry and other collective fruits, have so little connection with each other, and separate so completely at maturity as to partake more of the nature of a cluster than of a collective fruit. The same is true of the clump of tailed achenes that make up the fruit of the clematis. Though the product of a single flower, and thus techni-

186. — Head or cluster of dandelion fruits.

cally an aggregate fruit, they are really only a compact head or cluster. Some degree of cohesion is necessary to constitute a cluster of matured ovaries into an aggregate or a multiple fruit.

116. The Individual Fruits that make up the various kinds just described may belong to any of the classes mentioned in Sections 73–109; those of the blackberry, for instance, are drupes; of the strawberry, achenes; of the sweet gum, capsules.

117. Use of Fruits to the Plant. — Have you ever asked yourself how it could benefit a plant to invite birds and beasts to devour its fruit, as so many of the bright berries we find in the woods seem to do?

In order to answer this question we must remember that it is clearly to the advantage of every plant to disperse its

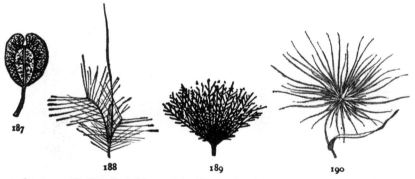

187–190. — Fruits adapted to wind dispersal: 187, winged pod of pennycress; 188, spikelet of broom sedge; 189, achene of Canada thistle; 190, head of rolling spinifex grass.

seeds as widely as possible, both that the seedlings may have plenty of elbow room, and that they may not have to draw their nourishment from soil already exhausted by their parents. The farmer recognizes this principle in the rotation of crops, because he knows that successive growths of the same plant will soon exhaust the soil of substances proper for its nutrition, while they may leave it rich in nourishment suitable for a different crop. Now, Nature, like a good farmer, seeks to provide for a rotation in her crops by furnishing all sorts of devices for the widest possible distribution of seeds. In the case of fleshy fruits this object is accomplished, for the most part, through the agency of animals. Our cultivated fruits have been so altered by man, and the parts useful to himself developed so exclusively for his own benefit that we cannot always judge from them exactly what service any particular organ would render to the plant. But in a state of nature, where the struggle for existence is so severe, no species can afford to develop any organ or quality that is

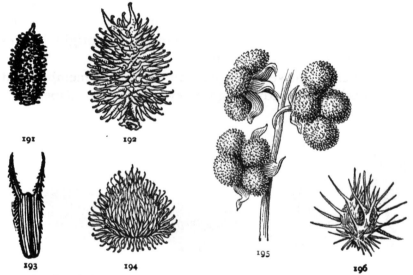

191–196.—Adhesive fruits: 191, pod of wild licorice; 192, cockle bur; 193, achene of bur marigold ; 194, burdock bur; 195, fruit of hound's tongue; 196, fruit of bur grass (*Cenchrus*).

not useful to itself. Hence, if you will examine the wild fruits of your neighborhood you will find that the edible ones generally produce hard, bony seeds, either too small to be destroyed by chewing, and thus capable of passing uninjured through the digestive system of an animal; or if too large to be swallowed whole, compelling the animal, by their hardness, or by their disagreeable flavor, to reject them.

On the other hand, where the seeds themselves are edible or attractive, the fruits are armed in every possible way against the assaults of animals. The acidity or other disagreeable qualities of most unripe fruits — the persimmon for instance — insures them pretty effectually against being molested before they have had time to mature their seeds, while in nuts and other indehiscent fruits, the protection afforded by their bony pericarp is frequently reënforced during the growing season by such appendages as the bur of the chestnut and the astringent hulls of the walnut and hickory nut.

The adaptations for dispersal in dry fruits consist mainly of wings and sails, like those of the maple, ash, thistle, dandelion, etc., by which they are carried from place to place by the wind or water; or of hooks and adhesive hairs by means of which they attach themselves to the coats of animals or the clothing of men, who are thus made the involuntary, and often the unwilling agents of their dispersal.

PRACTICAL QUESTIONS

1. To what class of fruits would you refer an ear of corn? of wheat? a sycamore or a buttonwood ball? a hop? a raspberry? a bunch of bananas? a pine cone? the fruit of the tulip tree and umbrella tree? of the mallow? Indian turnip?

2. Tell the nature of the individual fruits that compose each.

3. Name some fruits that are adapted to be carried about by the wind; by water; by animals.

4. How is the watermelon fitted for seed dispersal? the squash? fig? hickory nut? huckleberry? pomegranate? maypop (*Passiflora incarnata*)? corn? wheat? oats?

5. Could the last three survive in their present form without the agency of man?

6. Name all the plants you can think of that bear samaras and winged fruits of any kind; are they, as a general thing, tall trees and shrubs, or low herbs?

7. Name all you can think of that bear adhesive fruits, like the tockle bur and beggar's ticks; are any of these tall trees or shrubs?

8. Give a reason for the difference.

9. Why is the dandelion one of the most widely distributed weeds in the world?

10. Why is it that appendages for protection and dispersal are connected with the pericarp in indehiscent fruits and with the seeds in the dehiscent kinds?

FIELD WORK

Study the various edible fruits of your neighborhood with regard to their means of dissemination and protection. Consider the object of the protective devices in each case, whether against heat, cold, moisture, animals, etc.

Compare wild with cultivated fruits and notice in what respects man has altered the latter for his own benefit. Note, for instance, the difference between cultivated apples and the wild crab, between the cultivated grains and wild grasses. Observe the great number of varieties of each kind in cultivation and try to account for it.

Notice the situations in which different kinds of fruits grow, whether hot, dry, moist, windy, or sheltered, etc., and how they are affected by their surroundings.

Notice what animals feed upon the different kinds, and whether their visits are harmful or beneficial. Consider in what respects the interests of the plant itself, the interests of man, and the interests of other animals may clash or coincide.

IV. SEEDS AND SEEDLINGS

MONOCOTYLEDONS AND POLYCOTYLEDONS

MATERIAL — Dry and soaked grains of corn and oats, or other grasses. Seed of pine; the cones should be gathered in September or October, and kept until needed.

118. Dissection of a Grain of Corn. — Examine a dry grain of corn on both faces. Sketch the grooved side, labeling the hard, yellowish outer portion, *endosperm*, the depression near the center, *embryo*, or *germ*.

Next take a grain that has been soaked for twenty-four to twenty-six hours. What changes do you see? How do you account for the swelling of the embryo? Remove the skin and observe its texture. Is it a pericarp, a seed coat, or both? (Sec. 91.) Sketch the grain with the coat removed, labeling the flat oval body embedded in the endosperm, *cotyledon*, the upper end of the little bud-like body embedded in

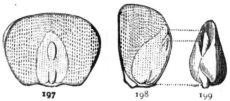

197   198   199

197-199. — Dissection of a grain of corn (GRAY): 197, soaked grain, seen flatwise, cut away a little and slightly enlarged, so as to show the embryo lying in the endosperm; 198, in profile section, dividing the grain through the embryo and cotyledon; 199, the embryo taken out whole. The thick mass is the cotyledon; the narrow body projecting upwards, the plumule; the short projection at the base, the hypocotyl.

the cotyledon, *plumule*, the lower part, *hypocotyl* — words meaning respectively, "seed leaf," "little bud," and "the part under the cotyledon." As this part has not yet differentiated into root and stem, we can not call it by either of these names. The cotyledon, the hypocotyl, and the plumule together compose the embryo. Pick out the embryo and sketch it as it appears under the lens, then

remove the plumule with the hypocotyl from the cotyledon, and sketch it. Make a vertical section of another
soaked grain at right anglés to its broader face, and sketch
it, labeling the parts as they appear in profile. Make a
cross section through the middle of another grain, and
sketch it. (A very sharp instrument must be used in
making sections, or they will not be satisfactory.)
What proportion of the grain is endosperm and what
embryo ?

It has been seen that one of the effects of iodine is to
turn starch blue, or even black (Sec. 26). Put a drop on
some of the endosperm and note the
effect. Of what does it consist?
Test the seed coats in the same way
to see if they contain any starch.

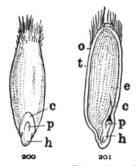

200, 201. — Dissection of
a grain of oats : 200, entire,
and slightly enlarged,
showing *c*, cotyledon, *p*,
plumule, *h*, hypocotyl; 201,
vertical section, *c*, cotyledon, *e*, endosperm, *p*,
plumule, *h*, hypocotyl.

**119. Study of a Typical Small
Grain.** — Make a similar examination
of a grain of oats or wheat. Compare
the endosperm of a soaked grain with
that of an unsoaked one; what change
has taken place and how do you
account for it ? Test with iodine and
see what it consists of. Which contains the greater proportion of endosperm, wheat (or oats) or corn ?

Notice that both the kinds of grain just examined have but
one cotyledon, hence, such seeds are said to be *monocotyledonous*. The grains are not typical seeds (Sec. 91),
but are selected for examination because they are large
and easy to obtain, and germinate readily. Other monocotyledonous seeds should be examined if practicable.
The blackberry lily (*Belamcanda*) and iris furnish good
examples.

120. Polycotyledons. — Remove one of the scales from
a pine cone and sketch the seed as it lies in its place on
the cone scale. The seed with its wing looks very much

like a samara of the maple, but it differs from all forms of the achene in being a true seed and not a fruit. Notice that the pine has no closed seed vessel, or ovary, like the other specimens we have been considering, but bears its seed naked in the axil of the cone scales, which may be considered open carpels. Hence, plants of this kind are called *Gymnosperms*, a word that means "naked seeds."

202, 203. — Pitch pine seeds (GRAY): 201, scale, or open carpel, with one seed in place; 203, winged seed, removed.

Look at the bottom, or little end of the seed, with your lens, for a small opening like a pin hole. Make an enlarged drawing of the seed as it appears under the lens, labeling this hole *micropyle*, a Greek word meaning "a

204. — Section of pine seed, showing the poly-cotyledonous embryo (GRAY).

little gate," because it is the entrance to the interior of the seed.

Remove the coat from a seed that has been soaked for twenty-four hours, and examine it with a lens. Pick out the embryo from the endosperm. Does the endosperm resemble that of the corn and wheat? Test it with iodine for starch. How does the embryo differ from those already examined? How many cotyledons are there?

Plants having more than two seed leaves are said to be *polycotyledonous*, a word meaning "having many cotyledons." This structure is characteristic of the pines, firs, hemlocks, and some other plants, mostly belonging to the Gymnosperms, or naked-seeded class.

PRACTICAL QUESTIONS

1. What gives to Indian corn its value as food? To oats; wheat; barley; rye; rice? (118, 119.)

2. Which of these grains have the larger proportion of starch or other endosperm to the embryo?

3. Do the husks or seed coats contain any nourishment?

4. Is there any nourishment in the embryos, apart from the endosperm?

5. What is bran?

6. Why will hogs fatten in a pine thicket in autumn?

DICOTYLEDONS

MATERIAL. — Dry and soaked seeds of the common bean, cotton, and castor bean. Where cotton can not be obtained, okra, maple, ash, morning-glory, or any other convenient specimens may be used, provided they are selected so as to show both the albuminous and the exalbuminous structure. Squash, pumpkin, horse-chestnut, etc., also make good studies. Beans should be put to soak from 12 to 24 hours before used; cotton about 48; squash and pumpkin from 3 to 5 days, and very hard seeds like the okra, castor bean, and morning-glory from 7 to 10. If such seeds are *clipped* before soaking, that is, if a small piece of the coat is chipped away from the end opposite the scar, they will soften more quickly. Keep them in a warm place with an even temperature till just before they begin to sprout, when the contents become softened. Very brittle cotyledons may be softened quickly by boiling them for a few minutes.

121. Examination of Some Typical Seeds. — Take a bean from the pod, noticing carefully its point of attachment. Lay it on one side and sketch it, then turn it over and draw the narrow edge that was attached to the pod. Notice the rather large scar (commonly called the eye of

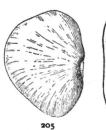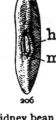

205, 206. — A kidney bean: 205, side view; 206, rhaphal view, showing *h*, hilum, *m*, micropyle.

the bean) where it broke away from the point of attachment. Label this in your drawing, *hilum.* Just below the hilum, look for a minute round pore like a pin hole. Label this *micropyle.* Compare a soaked bean with a dry one; what difference do you perceive? How do you account for the change in size and hardness? Find the hilum and the micropyle in the soaked bean. Make a section through the long diameter at right angles to the flat sides, press it slightly open and sketch it. Notice the line or slit that seems to cut the section in half longitudinally, and the small round object between the halves at one end ; can you tell what it is ?

Slip off the coat from a whole bean and notice its texture. Hold it up to the light and see if it shows any signs of veining. See whether the scar at the hilum extends through the kernel, or marks only the seed coat. Does the coat seem to adhere to the kernel more firmly at one point than another ? If so, label this point *chalaza*. Lay open the two flat bodies into which the kernel divides when stripped of its coats. Sketch their inner face and label them *cotyledons*. Be careful not to break or displace the tiny bud packed away between the cotyledons, just above the hilum. Label the round, stem-like portion of this bud, *hypocotyl*, and the upper, more expanded part, *plumule*. Which way does the base of the hypocotyl point, toward the micropyle, or away from it ? Pick out this budlike body entire and sketch as it appears under the lens. Open the plumule with a pin and examine it with a lens ; of what does it appear to consist ? Do you find any endosperm around the cotyledons as in the corn and oats ? Break one of the soaked cotyledons, apply some iodine to it, and report whether it contains any starch. Where is the nourishment for the young plant stored ? What part of the bean gives it its value as food ?

207. — Cotyledon of a bean, showing plumule.

Notice that in the bean the embryo consists of three parts, the hypocotyl, plumule, and cotyledons, which completely fill the seed coats, leaving no place for endosperm.

208. — Cotton seed with lint.

122. Dissection of a Cotton Seed. — (Where cotton can not be obtained, morning-glory, okra, or maple may be used.) Scrape the lint from a seed of cotton as closely as possible, or if practicable, get a specimen of one of the smooth seeded varieties in cultivation, and look for a faint line or groove on one side, leading from the small end to the big end. Make a sketch of the side showing this line, label it *rhaphe*, and the point where it begins, at the large end of the seed, *chalaza*. Look for the hilum at the other

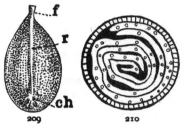

end of the rhaphe, and for the micropyle near it, at the small end of the seed. If they can not be distinguished on account of the lint, make a longitudinal section of a well-soaked seed and find where the hypocotyl points. Which way did it point in the bean? This is the case with all seeds; the base of the hypocotyl is towards the micropyle, and so we can

209, 210. — Dissected cotton seed: 209, seed with lint removed (magnified three times), *f,* funiculus, or seed stalk, *r*, rhaphe, *ch*, chalaza; 210, cross section of the seed still more highly magnified, showing the crumpled cotyledons.

always tell where the micropyle is by noticing which way the hypocotyl points. Make an enlarged sketch of the section as it appears under the lens, and also of a cross section of another soaked seed about midway between the two ends, showing as accurately as you can the lines of any folds or convolutions that you may see. Label such parts as you can clearly make out, leaving the others till after further examination.

From a seed that has been boiled for five or ten minutes to soften the contents, gently remove the coats so as to leave the embryo whole. How many seed coats are there? How do they differ in color and texture? Try to distinguish them in the sketches you have made, and label the hard outer one that corresponds to the shell of an egg, *testa*, the soft inner one, *tegmen*. What is the use of each? As the coats were removed did

211. — Embryo with cotyledons partly unfolded.

they seem to adhere to the kernel more tenaciously at one point than elsewhere? Look for a little dark spot inside near the base, that marks where the seed coats and kernel adhered together. Refer to your sketch of the outside of the seed, and say to what it corresponds. Are the chalaza and micropyle close together, as in the bean, or at opposite ends of the seed?

Sketch the kernel, or embryo, without opening it, as it appears under the lens. Notice the irregular fold or groove down one side that divides it into two nearly equal parts. Label these *cotyledons*. Observe the complicated way in which they are folded. Try to imitate it with a piece of paper. Would any other way of folding fit them so snugly into the seed coats? Straighten them out as well as you can and sketch them. Which are most leaf-like, the cotyledons of the bean or the cotton? Are either of them at all similar in shape to the foliage leaves of their respective plants? How do they compare in size relatively to the size of the respective seeds? Which are best fitted to perform the office of true leaves?

In seeds like the pea and bean, where the cotyledons are too thick and clumsy to do well the work of true leaves the young plant will need a well-developed plumule to begin life with, but where the cotyledons are thin and leaflike, as in the cotton, and to a less degree in the pumpkin and squash, and capable of developing quickly into true leaves, there is generally no plumule formed in the embryo.

123. The Castor Bean. — Lay a castor bean on a sheet of paper before you with its flat side down; what does it look like? The resemblance may be increased by soaking the seed a few minutes, in order to swell the two little protuberances at the small end. Can you think of any benefit a plant might derive from this curious resemblance of its seed to an insect?

Sketch the seed as it lies before you, labeling the protuberance at the apex, *caruncle*. The caruncle is no essential part of the seed, but a mere appendage developed by various plants, the use of which is not always clear. What appears to be its object in the castor bean? It may occur on any part of the seed, though it generally takes some other name when borne elsewhere than at the micropyle, of which it is usually an outgrowth.

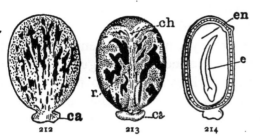

212–214. — Castor bean (slightly magnified): 212, back view; 213, front view, *ch*, chalaza, *r*, rhaphe, *ca*, caruncle; 214, vertical section, *e*, embryo, *en*, endosperm.

Turn the seed over and sketch the other side. Notice the colored line or stripe that runs from the large end to the caruncle. This represents the rhaphe. Its starting point near the large end, which is marked in fresh seeds by a slight roughness, is the chalaza. Where the rhaphe ends, just at the beak of the caruncle, you will find the hilum. The micropyle is covered by the caruncle, which is an outgrowth from it.

Next cut a vertical section through a seed that has been soaked for several days, at right angles to the broad sides, and sketch it. Label the thick outer coat *testa*, the delicate inner one *tegmen*, the white, pasty mass within that, *endosperm*. Can you make out what the narrow white line running through the center of the endosperm, dividing it into two halves, represents? Make a similar sketch of a cross section. Notice the same white line running horizontally across the endosperm, dividing it into two equal parts. To find out what these lines are, take another seed (always use soaked seeds for dissection) and remove the coats without injuring the kernel. Notice the little dark spot where it was joined to the coats at the chalaza. Split the kernel carefully round the edges, remove half the endosperm, and sketch the other half with the delicate embryo lying on its inner face. You will have no difficulty now in recognizing the lines in your drawings as sections of the thin cotyledons. Where is the hypocotyl, and which way does its base point? Remove the embryo from the endosperm, separate the cotyledons with a pin, and hold them up to the light to see their beautiful texture.

Sketch them under the lens, showing the delicate venation. Is there any plumule?

Test the endosperm with a little iodine to see if it contains any starch. Crush a bit of it on a piece of white paper and see if it leaves a grease spot. What does this show that it contains? Test the embryo in the same way, and see whether it contains any oil.

124. Arrangement of the Embryo. —Notice the difference in the way the embryo is packed in the castor bean, and in such seeds as the cotton, okra, and maple. In the former it is said to be *straight*, while in the latter it is

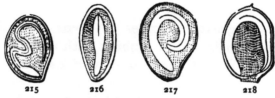

215–218. —Arrangement of embryo in endosperm (GRAY): 215, morning-glory; 216, barberry; 217, potato; 218, four o'clock.

folded or *plicate*. In different seeds it may be coiled and folded in many different ways. It may also be packed within the endosperm, as the castor embryo, or coiled or wrapped around it, as in the chickweed.

125. Storage of Nourishment in the Seed. — In the various seeds examined we have seen that the nourishment for the young plant is either stored in the embryo itself, as in the cotyledons of the bean, acorn, squash, etc., or packed about them in the form of endosperm, as in the corn, wheat, and castor bean.

The latter are classed by botanists as *albuminous*, the former as *ex-albuminous*—the word "albumin" referring not to the chemical composition of the food supply, but to its office, which is similar to that of the albumen, or white of the egg stored up for the nourishment of the hatching chick. The older botanists, recognizing the analogies between the seed and the egg, and not understanding the true nature of either, regarded the seed as a sort of vegetable egg, and named the reserve material we now call endosperm, albumen. It is now known to be something very different, however, from the white of an egg. Frequently it is starch, as we have seen in the corn, wheat, oats, etc., or it may be an oil, as in the castor bean and peanut, or something quite different from either. Hence, modern botanists have renamed this substance endosperm, a word meaning merely something contained "within the seed," and therefore applicable to any kind of substance. The old adjectives, albuminous and ex-albuminous, have been retained for want of something better — ex-endospermous being such an awkward compound that even botanists hesitate to use it.

By far the greater number of seeds are albuminous; that is, they consist of an embryo with more or less nourishing matter stored about it in various ways. Even in ex-albuminous seeds the endosperm is present; it has merely been absorbed and stored in the cotyledons before germination.

126. Principles of Classification. — We are now prepared to understand the great fundamental distinctions upon which botanists base their classification of *Spermatophytes*, or seed-bearing plants. The first division depends upon the presence or absence of a seed vessel, and ranges all the higher plants into two classes according to this feature. The first division embraces the

127. Gymnosperms, or naked-seeded plants, of which we have had an example in the pine. They are the most primitive type of seed-bearing plants and the most ancient. Though they are not so abundant now as in past ages, numbering only about four hundred known species, they present many diversities of form, which seem to ally them on the one hand with the lower, or spore-bearing plants (ferns, mosses, etc.), and on the other with the

128. Angiosperms, or plants that produce their seeds in a special covering of closed carpels, like most of the fruits and pods that we have been considering. This group contains all the true flowering plants, and forms the most important part of the vegetation of our globe, numbering not less than one hundred thousand species. It is divided into two great groups, distributed, as we have seen, according to the number of their cotyledons, into

129. Monocotyledons and Dicotyledons. — These are further distinguished by the fact that dicotyledons have, as a general thing, net-veined, and monocotyledons, parallel-veined leaves. The cause of this difference, science has never yet been able to explain, so that for the present we shall have to accept it as a fact which we can not understand. There are other differences, also, in the structure of the flower and the stem, which will be considered later.

PRACTICAL QUESTIONS

1. Make a list of all the seeds you can think of that have very thick cotyledons.

2. Draw a line under all that are used as food by man or beast.

3. Could a species derive any advantage from tempting animals to eat and destroy its seed? (117.)

4. What then is the advantage to the plant of providing this food supply? (125.)

5. Do you find any edible seeds without protection, and if so, account for their want of it.

6. Make a list of all the albuminous seeds you can think of that are used for food or other purposes, such as medicines and unguents.

7. Do you find as many food materials among these as among the ex-albuminous kind?

8. Are they in general as well protected as ex-albuminous seeds?

9. How do the two compare, in a general way, as to size?

10. What part of the following plants do we eat, the fruit or the seed? Corn; wheat; hickory nut; cocoanut; Brazil nut; peanut; beechnut; string beans; honey locust; coffee; anise; celery.

11. From what part of the castor bean do we get the oil? Of the peanut? Of the cotton seed?

12. What gives to cotton-seed meal its value as cattle food?

13. Is there any valid objection to the wholesomeness of peanut oil, and cotton-seed lard?

FORMS AND GROWTH OF SEED

MATERIAL. — Various kinds of pods and fruits with the seed still attached to the placentas, such as the following. Straight seeds: buckwheat, smilax, dock, knotweed.   Inverted: castor bean, cotton, violet, magnolia, cherry, apple, and the majority of common seeds. Curved: bean, purslane, jimson weed, okra, and most of the pink family.

130. Straight Seeds. — The most natural, and at the same time the least common mode of attachment is for a seed to

stand erect upon its stalk like a pink or a rosebud on its stem.   A seed that grows in this manner is said to be *orthotropous* (Figs. 220, 221).   If we imagine the seed coats to be separated from each other and from the em-

219. — An erect flower, showing attachment of the stalk.

bryos, as in the diagram (Fig. 220), we shall see that the parts all come together and coalesce at the base, where they are attached to the seed stalk, just as all the parts of a flower adhere at the receptacle (Fig. 221).   This point,

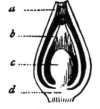

220. — Diagrammatic section of a typical or orthotropous seed (GRAY), showing the outer coat, *a*; the inner, *b*; the nucleus, *c*; the chalaza, or place of junction of these parts, *d*.

the organic base of the seed, is the chalaza, and

you can now understand the tendency of the coats of the different seeds examined to cohere there.   An inspection of the diagram will show that in orthotropous seeds the hilum and chalaza will always coincide.   At the other end, the tip or apex of the seed (Fig. 220), the coats do not quite come together, thus causing the

221. — Section of an upright flower, showing insertion of parts at base (*after* GRAY).

small aperture that we labeled "micropyle" in our drawings.   In this arrangement the micropyle will always be opposite the chalaza, and it marks the organic apex of the seed as the chalaza does its base.

131. Inverted Seeds. — But sometimes a flower turns over on its stalk, like the snowdrop and harebell, and the same thing often happens to a seed. This gives rise to the inverted, or *anatropous* kind (Fig. 223). In this case, which is due to certain peculiarities in the early growth of the seed, the stalk does not remain separate like the stem of a pendent flower, but coalesces more or less completely with the coats, and thus forms the rhaphe (Fig. 223), *d*. The chalaza remains at the base, *ch*, which is now by inversion at the top; but as the stalk, or rhaphe, is adherent to the coats, it can not break away at the base, and

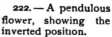

222. — A pendulous flower, showing the inverted position.

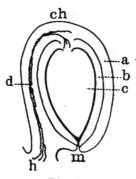

223. — Diagram of an inverted or anatropous seed, showing the parts in section: *a*, outer coat; *b*, inner coat; *c*, nucleus; *d*, rhaphe; *ch*, chalaza; *h*, hilum; *m*, micropyle (*after* GRAY).

hence, in anatropous seeds the hilum and micropyle are brought close together, at the real apex of the seed. The adherent stalk, or rhaphe, often becomes reduced to a mere line or groove, as we saw in the cotton and castor bean, or may disappear altogether, but the chalaza can generally be distinguished by a tendency of the parts to cohere at that point.

Variations in these modes of attachment are shown in Figures 225, 226. In the campylotropous or curved kind, the seed is bent over during early growth into a circular or kidney shape, so that the micropyle is brought into close

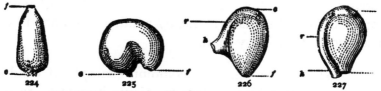

224-227. — Seeds (GRAY): 224, orthotropous seed of buckwheat, *c*, hilum and chalaza, *f*, micropyle; 225, campylotropous seed of a chickweed, *c*, hilum and chalaza, *f*, micropyle; 226, amphitropous seed of mallow, *f*, micropyle, *h*, hilum, *r*, rhaphe, *c*, chalaza; 227, anatropous seed of a violet, the parts lettered as in the last.

juxtaposition with the hilum, as we saw in the bean. **How**
does this differ from the anatropous kind? Compare the
seed you have examined and the drawings you have made
with Figures 224–227, and see if you can tell to which
class each belongs. Why are these distinctions not applicable to corn and other grains? (Sec. 91).

132. Position in the Pericarp. — The terms "orthotropous," "anatropous," etc., refer to the position of the seed
on its footstalk and have nothing to do with its attachment

228–230. — Position of seeds in the carpels: 228, erect seed of Ceanothus; 229,
horizontal anatropous seeds of the European star-of-Bethlehem; 230, suspended
seeds of Polygala.

to the pericarp, which may be either erect, horizontal, or
suspended. An orthotropous seed may hang bottom upwards from the apex of the carpel without altering its
character ; and in like manner one of the anatropous kind
may be attached in such a way as to bring it back, by a
double inversion, to the upright position. The castor bean
furnishes a good example of this.

133. Seed Dispersal. — This subject has already been
touched upon in the chapter on fruits, and the object of
distribution is in both cases the same. The agencies of
dispersal are either natural, *i.e.* by wind, water, and animals,
or artificial, that is, by man.

134. Wind Dispersal. — A common example of wind dispersal is afforded by the class of plants known to farmers
as "tumble weeds." Well-known examples of these are the
Russian thistle, winged pigweed, "old witch grass," hair

grass, etc.   Such plants generally grow in light soils and either have very light root systems, or are easily broken from

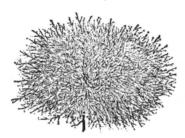

231.—A fruiting plant of winged pigweed (*Cycloloma*), showing the bunchy top and weak anchorage of a typical tumble weed.

232.—Panicle of "old witch grass," a common tumble weed.

their anchorage and left to drift about on the ground. The spreading, bushy tops become very light after fruiting so as to be easily blown about by the wind, dropping their seeds as they go, until they finally get stranded in ditches and fence corners, where they often accumulate in great numbers during the autumn and winter.

In the Japan varnish tree (*Sterculia platanifolia*) the seeds remain attached all through the winter to the open follicle, which becomes very light when dry, and acts as a sort of float for wafting the seeds away on every breeze.

135. Explosive Capsules. — Some plants undertake to disperse their seeds without the intervention of any external

233   234   235

233–235. — 233, A pod of wild vetch, with mature valves twisting spirally to discharge the seed; 234, pod of crane's-bill discharging its seed; 235, capsules of witch-hazel exploding.

agent. Examples of this kind are the violet, witch-hazel, and touch-me-not, whose capsules dehisce with a little explosion and shoot out the seeds as if they were fairy mortars. It is worth while to gather a bough of witch-hazel in winter and keep it in the schoolroom to watch the explosions. In other cases, the carpels curl upwards with a sudden jerk, as in some of the geranium family, or twist themselves into a spiral, like the valves of the rabbit pea (*Vicia · americana*), thus acting as a spring to eject the seeds.

136. Animal Agency. — Examples of adaptation for dispersal by means of animals were given in Section 117, but by far the most active agent in the dissemination of both fruits and seeds is man. This is the frequent result of intention on his part, in the introduction and cultivation of new grains, fruits, and vegetables, and he works to the same end unconsciously and often to his great detriment by the transportation of the bulbs or seeds of pernicious weeds in the dirt clinging to hoes and plowshares, and the mixture of impurities with his crop seeds through ignorance, carelessness, or unavoidable causes. This mode of dispersal, however, is purely artificial, and except in the case of a few weeds that have adjusted themselves to the conditions of cultivation, is not correlated with any special adaptations in the plants themselves, many of our most widely distributed weeds, such as the rib grass, or common plantain, the mayweed and the narrow-leaved sneezeweed, possessing very imperfect natural means of dispersal.

PRACTICAL QUESTIONS

1. Name the ten most troublesome weeds of your neighborhood.
2. What natural means of dispersal have they?
3. Which of them seem to owe their propagation to man?
4. Are there any tumble weeds in your neighborhood?
5. Should you expect to find such weeds abundant in a hilly or a very woody country?
6. What situations are best fitted for their propagation?
7. Make a list of all the seeds you can think of that are adapted to dispersion by the wind; by water; by animals.

8. Mention some of the ways in which weeds can be propagated by careless farmers.

9. Why are so many strange weeds or other new plants found first along railroad tracks?

10. Account for the absence of weeds in forests and groves.

11. Suggest ways for checking the propagation of weeds, and of stopping their introduction.

GERMINATION

MATERIAL. — Seed of any kind that will germinate readily and with a moderate degree of heat. Corn, oats, cotton, beans, mustard, will any of them answer. Six or eight ordinary preserving jars, or bottles. Some moist cotton, sawdust, or layers of blotting paper, or old flannel. Some vaseline, or, if this is not at hand, lard.

137. Conditions of Germination. — If kept perfectly dry, seed may sometimes be preserved for months, or even years. Peas have been known to sprout after ten years, red clover after twelve, and tobacco after twenty. Ordinarily, however, the vitality of seeds diminishes with age, and in making experiments it is best to select fresh ones. The ones used for comparison should also, as far as possible, be of the same size and weight.

138. Moisture. — Can seeds have too much moisture? To answer this question drop a number of dry grains of corn, oats, or other convenient seed, into a bottle or other vessel with a bedding of cotton or paper that is barely moistened, and an equal number of soaked seeds of the same kind into another vessel with a saturated bedding of the same material. In a third vessel place the same number of soaked seed, covering them partially with water, and in a fourth cover the same number entirely. Label them 1, 2, 3, and 4, and keep all together in a warm and even temperature, and note the rate of germination in the different vessels.

139. Air. — Next arrange in a similar manner a glass jar containing the same kind of seed as before, using a sufficient quantity to fill it at least half full. The vessel should be large enough to hold at least a liter (about one quart). Seal it hermetically so as to prevent the access of fresh air. Label it 5, and place it with the other four. The water used for soaking the seeds and for moistening the bedding in this experiment should first have had its contained air expelled by boiling.

To test the behavior of seeds in the entire absence of air is difficult, because it is not possible to expel all traces of the atmosphere even with an air pump.

140. Temperature. — Arrange some soaked seeds in three or four different vessels just as in No. 2, in the first experiment, and place where they will be subjected to different temperatures, ranging say from 0° to 30° C. (about 32° to 86° F.). Test frequently with a thermometer, keeping the temperature as even as possible, and maintaining an equal quantity of moisture in each vessel. Keep a record of the number of seed sprouted in each after every twenty-four hours. In those parts of the South where the cold is not continuous enough to keep seed from germinating under ordinary conditions, experiments in low temperatures can not very well be made unless there is a refrigerator available. In sections where there is continuous cold, tests might also be made of the minimum temperature at which different seeds will germinate. Sachs found the minimum for corn to be 9°.4 C. (about 49° F.), and for the gourd, 14° C. (about 58° F.).

141. Recording Observations. — Arrange a page of your notebook after the model given below, and record your

NUMBER OF SEEDS GERMINATED

		24	48	72	4 d.	5 d.	6 d.	7 d.	8 d.	10 d.	w 2.
No. of hours . .											
No. of vessel . .	1										
No. of vessel . .	2										
No. of vessel . .	3										
No. of vessel . .	4										
No. of vessel . .	5										
No. of vessel . .	6										

observations at intervals of twenty-four hours. When most of the seeds in jar 5 (Sec. 139) have begun to sprout, insert a thermometer and let it remain two or three minutes. Does it indicate any change of temperature? Refer to Section 29 and account for the change. If cotton seed are used, the rise of temperature will be very marked.

142. Vitality. — A very interesting point is to test the temperature at which different seed lose their vitality, by subjecting dry and soaked ones of various kinds to different degrees of heat and cold. Notice how the extremes tolerated are affected by: first, the length of time the seeds are exposed; second, by the amount of water contained in them; and third, by the nature of the seed coats. Every farmer knows that the effect of freezing is much more injurious to plants or parts of plants when full of sap (water) than when dry. This is because in freezing the water expands and ruptures the tissues, thus setting up internal disturbances which are liable to result in death, especially if thawing takes place so rapidly that the life processes have not time to readjust themselves. In like manner it will be found that when seeds are subjected to moist heat, they are killed at a lower temperature and in a shorter time than when dry. When heated in water of the same temperature, those seeds will be found to resist best whose coats are most impervious to the liquid.

143. Time Required for Germination. — Arrange in a bed of moist sand, placed between two soup plates, seeds of various kinds. Good specimens would be some of the following: corn, wheat, peas, cotton, okra, turnip, apple, morning-glory, orange, grape, persimmon, castor bean, peanut, etc. Clip some of the harder ones and place them in the same germinator with unclipped ones. Keep all under similar conditions as to temperature, moisture, etc., and record the time required for each to sprout. What is the effect of clipping, and why?

144. The Relative Value of Perfect and Inferior Seed. —
From a number of seeds of the same species select half a

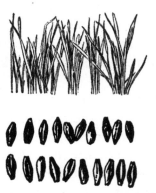

236. — Stem development of seedlings raised from healthy grains of
barley; weight, 39.5 grams (about
500 grs.).

237. — Stem development of seedlings
raised under exactly similar conditions
from the same number of inferior grains;
weight, 23 grams (about 350 grs.).

dozen of the largest, heaviest, and most perfect, and an
equal number of small, inferior ones. If a pair of scales
is at hand, the different sets should be weighed and a

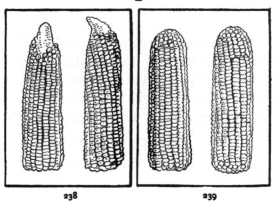

238   239

238, 239. — Improvement of corn by selection: 238,
original type; 239, improved type developed from it.

record kept for
comparison with
the seedlings at
the end of the ex-
periment. Plant
the two sets in
pots containing
exactly the same
kind of soil, and
keep under identi-
cal conditions as
to light, tempera-
ture, and moisture.

Keep the seedlings under observation for two or three
weeks, making daily observations and occasional drawings
of the height and size of the stems, and the number of
leaves produced by each.

These experiments can be carried on simultaneously
with the study of *Seedlings* and *Growth*. It is not

expected that any one class will have time to complete them all, but a number are suggested in order that different teachers may choose the ones best suited to their circumstances.

PRACTICAL QUESTIONS

1. What are the principal external conditions that affect germination? (137, 138, 139, 140.)

2. What effect has cold? Want of air? Too much water?

3. Is light necessary to germination?

4. What is the use of clipping seeds?

5. In what cases should it be resorted to?

6. Why will seed not germinate in hard, sun-baked land without abundant tillage? Why not on undrained or badly drained land? (138, 139.)

7. Will seeds that have lost their vitality swell when soaked?

8. Are there any grounds for the statement that the seeds of plums boiled into jam have sometimes been known to germinate?[1] (142.)

9. Could such a thing happen in the case of apples or watermelons, and why or why not? (142.)

10. Does it make any difference in the health and vigor of a plant whether it is grown from a large and well-developed seed or from a weak and puny one? (144.)

11. Would a farmer be wise who should market all his best grain and keep only the inferior for seed?

12. What would be the result of repeated plantings from the worst seed?

13. Of constantly replanting the best and most vigorous?

SEEDLINGS

MATERIAL. — Seedlings of various kinds in different stages of growth. Those from seeds experimented with in Sections 137–144 may be used to begin with. Corn, oats, bean, squash, cotton, are the ones mentioned in the text. Ash, maple, morning-glory, or castor bean may be used instead of cotton, but the last two are rather difficult to germinate, requiring from 8 to 10 days, or even longer, if the temperature is too low. Soaked seeds of cotton and corn will germinate in from 3 to 7 days, according to the temperature; oats in 1 to 4, beans in 4 to 6, squash in 8 to 10. Germination will be greatly facilitated by soaking the seeds for 12 to 24 hours before planting them, and very obdurate ones may be forced by clipping.

[1] Vines, "Lectures on the Physiology of Plants," p. 282. See also, Sachs, "Physiology of Plants."

A good germinator can be made by putting moist sand or sawdust between two plates. The germinator should be kept at an even temperature of about 20° C. (70° F.). Seeds even of the same kind develop at such different rates that it will probably not be necessary to make more than two plantings of each sort, about 4 or 5 days apart. Enough must be provided to give each pupil 3 or 4 specimens in different stages of development.

145. Seedlings of Monocotyledons. — Examine a grain of corn that has just begun to sprout; from which side does the seedling spring, the plain or the grooved one? Refer to your sketch of the dry grain and see if this agrees with the position of the embryo as observed in the seed. Make sketches of four or five seedlings in different stages of advancement, until you reach one with a well-developed blade. Examine each carefully with regard to the cotyledon, the root, and the plumule. Which part first appeared above the ground? In what direction does the plumule grow? The hypocotyl? Does the cotyledon appear above ground at all? Slip off the seed coats and see if there is any difference in the size and appearance of the contents as you proceed from the younger to the older plants. How would you account for the difference?

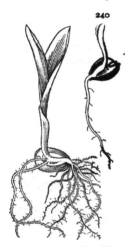

240, 241. — Seedling of corn (*after* GRAY): 240, early stage of germination; 241, later stage.

146. The Cotyledon. — Is the cotyledon of any use to the seedling when it remains in the ground? In order to answer this question, cut away carefully, so as not to injure the plumule, the cotyledon with its endosperm, from a very young seedling, and place on a piece of coarse netting stretched over a glass of water so that its roots will touch the liquid. Put beside it another seedling of the same age and size from which the cotyledon has not been removed, and watch their growth for a week or ten days. Which has developed most rapidly in that time? Test the cotyledon on the second seedling for starch; what has become of it? Test sections of the root and stem of the same seedling and see if any of the starch has gone into them.

147. Growth of the Plumule. — What part of the plumule comes out of the ground first? Is it straight or bent? Open the outer sheath of a well-developed plumule with a needle; what do you find inside? Examine the plumule of an older plant that has developed several leaves; where does the second one come from? Look within that for the next one; from where does the new leaf always seem to proceed? Measure the internodes from day to day and note their rate of growth in your book.

148. Growth of the Root. — Examine the lower end of the hypocotyl and find where the roots originate. Observe their tendency to spread out in every direction, and even to develop from the lower nodes of the hypocotyl; would you say that the roots are an outgrowth from the stem, or the stem from the root? Mark off a root into sections by moistening a piece of sewing thread with indelible ink and applying it to the surface of the root at intervals of about one millimeter ($\frac{1}{20}$ of an inch). Lay the seedling on a moist bedding in a glass jar, covered lightly to prevent evaporation, and watch to see in what part of the root growth takes place.

242, 243. — Seedling of corn, marked to show region of growth: 242, early stage of germination; 243, later stage.

Notice the grains of sand or sawdust that cling to the rootlets of plants grown in a bedding of that kind. Examine with a lens and see if you can account for their presence. Lay the root in water on a bit of glass, hold up to the light and look for root hairs; on what part are they most abundant?

149. Root Hairs are the chief agents in absorbing moisture from the soil. They do not last very long, but are

244. — Seedling of wheat, with root hairs.

constantly dying and being formed again in the younger and tenderer parts of the root. These are usually broken away in tearing the roots from the soil, so that it is not easy to detect them except in seedlings, even with a microscope. In oat and maple seedlings they are very abundant and clearly visible to the naked eye. The amount of absorbing surface on a root is greatly increased by the presence of the hairs; and they exude, moreover, a slightly acid secretion, which aids them in dissolving and absorbing the mineral substances contained in the particles of earth and sand to which they adhere.

150. The Root Cap. — Look at the tip of the root through your lens and notice the soft, transparent, crescent or horseshoe-shaped mass in which it terminates. This is the root cap and serves to protect the tender parts behind it as the roots burrow their way through the soil. Being soft and yielding, it is not so likely to be injured by the hard substances with which it comes in contact as the more compact tissue of the roots. It is composed of the loose cells out of which the solid root substance is being formed, and the growing point of the root is at the extremity of the tip just behind the cap (Fig. 245). The cap is very apparent in a seedling of corn, and can easily be seen with the naked eye, especially if a thin longitudinal section is made. It is also well seen in the water roots of the common duckweed (*Lemna*), and on those developed

245. — Magnified section of root tip: *c*, root cap; *g*, growing point.

by a cutting of the wandering Jew, when placed in water. Are there any hairs on the root cap?

A good way to study the small, delicate parts of plants is to place them between two thin, clear pieces of glass and hold up to the light. Even without a lens·many peculiarities of structure can in this way be made apparent to the eye.

Instead of corn, seedlings of wheat or oats may be used, and if time permits it would be well to examine and compare the two.

151. Organs of Vegetation. — These three organs, root, stem, and leaf, are all that are necessary to the individual life of the plant. They are called organs of vegetation in contradistinction to the flower and fruit, which constitute the organs of reproduction. The former serve to maintain the plant's individual existence, the latter to produce seed for the propagation of the species, so we find that the seed is both the beginning and the end of vegetable life.

152. Polycotyledons. — The pine is very difficult to germinate, requiring usually from 18 to 21 days, but if a seedling can be obtained it will make an interesting study. By soaking the mast for 24 hours and planting in damp sand kept at an even temperature of not less than 23° C. (74° or 75° F.) a few specimens may be obtained.

246. — Seedling of pine (GRAY).

153. Seedlings of Dicotyledons. — Sketch, without removing it, a bean seedling that has just begun to show itself above ground; what part is it that protrudes first? Sketch in succession four or five others in different stages of advancement. Notice how the hypocotyl is arched where it breaks through the soil. Can you account for this? Does it occur in the monocotyledons examined? Almost all dicotyledons exhibit this peculiarity in germination; can you see what causes it? Do the cotyledons appear above ground? How do they get out? Can you perceive any advantage in their being dragged out of·the ground backwards in this way rather than

247. — Seedlings of bean in different stages of growth: *cc*, cotyledons, showing the plumule and hypocotyl before germination; *a, b, d*, and *e*, successive stages of advancement. At *d* the arch of the hypocotyl is beginning to straighten; at *e* it has entirely erected itself.

pushed up tip foremost? What changes have the cotyledons undergone in the. successive seedlings? Remove from the earth a seedling just beginning to sprout and sketch it. From what point does the hypocotyl protrude through the coats? Does this agree with its position as sketched in your study of the seed? In which part of the embryo does the first growth seem to have taken place?

Remove in succession the several seedlings you have sketched and note their changes. How does the root differ from that of the corn and oats? Look for root hairs; if there are any, where do they occur? Mark off the root of a young seedling into sections as directed in Section 148, and

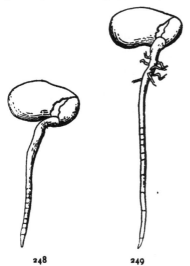

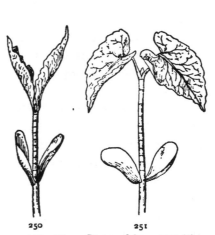

248   249
248, 249. — Root of bean seedling, measured to show region of growth: 248, early stage of germination; 249, later stage.

250, 251
250, 251. — Stem of bean seedling, measured to show region of growth: 250, early stage of growth; 251, later stage.

watch it from day to day. In what part does growth take place? Mark off a node of the stem in a similar manner and find out how it grows. Allow a seedling to develop until it has put forth several leaves, and measure daily the successive internodes. Does an internode stop growing when the one next above it has formed? When is growth most rapid? Reverse the position of a number of seedlings that have just begun to sprout and watch what will happen. A good way to observe the growth of roots is to fill a glass jar or a lamp chimney with moist cotton or sawdust, and insert the seedling between the side of the jar and the moist filling.

154. Cotton. — Examine a number of cotton seedlings in different stages of growth. What part appears above ground first? How does this compare with the first appearance of the bean? Of corn and oats? Pull up a seedling that has just begun to sprout; does the root come from the big or the little end of the seed? Does this agree with what you learned about the position of the hypocotyl in Sections 121 and 122? Notice how the coats adhere at the chalaza, even after the cotyledons are well above ground; is this woolly nightcap of any special service to a delicate plant like the cotton? Notice the little speckled glands that cover the stem and the cotyledons. What change of color do the latter undergo as the seedling develops? How do they compare as foliage leaves with those of the bean, squash, etc.? With the foliage leaves of the mature cotton plant? Of what use to a plant are the cotyledons when they appear above ground? To answer this question cut away the cotyledons from a number of seedlings as soon as they appear, and observe the result as compared with others that have not been cut.

Pull up a seedling and sketch it entire, showing the long, straight taproot. How does it compare in length with that of the bean? How do both differ from those of the corn and oats? Measure the growth of the root and stem as you did in the bean. Reverse the position of a number of seedlings, so that the hypocotyl shall point upward and the plumule downward, and watch the effect upon their growth. After a few days reverse them again and note the effect. In sections where cotton seed can not be obtained, maple, ash, morning-glory, or squash, pumpkin, etc., may be substituted.

PRACTICAL QUESTIONS

1. Do the cotyledons, as a general thing, resemble the mature leaves of the same plants?

2. Try to account for the difference, if you observe any; could convenience of packing in the seed coats, for instance, have anything to do with it?

3. If seeds are planted in the ground in a number of different positions, will there be any difference in the position of the seedlings as they appear above ground?

4. Of what advantage to the farmer is this tendency of seedlings to right themselves?

GROWTH

MATERIAL. — A flower pot suspended by a wire, some bulbs, and several well-developed seedlings to experiment with.

155. What Growth Is. — With the seedling begins the growth of the plant. Most people understand by this word, mere increase in size; but growth is something more than this. It involves a change of form, usually, but not necessarily, accompanied by increase in bulk. Mere mechanical change is not growth, as when we bend or stretch an organ by force, though if it can be kept in the altered position till such position becomes permanent, or as we say in common speech, " till it grows that way," the change may become growth. To constitute true growth, the change of form must be permanent, and brought about, or maintained by forces within the plant itself.

Remove the scales from a white-lily bulb, weigh them, and lay them in a warm, not too damp, place, away from light.

After a time young bulblets will form at the base of each scale. Weigh the scales again, and if there has been any loss, account for it (see Sections 24–27, and 65). The same experiment can be tried by allowing hyacinth or other bulbs to germinate without absorbing moisture enough to affect their weight.

156. Conditions of Growth. — The internal conditions depend upon the organization of the plant. The essential external conditions are : food material, water, oxygen, and a sufficient degree of warmth. It may be greatly influenced by other circumstances, such as light, gravitation, pressure, and (probably) electricity, but the four first named are the essential conditions without which no growth is possible.

157. Region of Growth. — It was seen in Sections 148 and 153 that the region of active growth in the root is just above the tip, behind the cap. In the stem the region of increase is more evenly distributed, the lower nodes continuing to grow for some time after the others are formed, but a little observation will show that in stems also, growth is usually most active in the region near the apex, where new cells are being produced.

158. Cycle of Growth. — When an organ becomes rigid and its form fixed, there is no further growth, but only nutrition and repair, processes which must not be confounded with it. Every plant and part of a plant has its period of beginning, maximum, decline, and cessation of growth. The cycle may extend over a few hours, as in some of the fungi, or, in the case of large trees, over thousands of years.

159. Direction of Growth. — Plant in a pot suspended as shown in Figure 252, a healthy seedling of some kind, two or three inches high, so that the plumule shall point downward through the drain hole and the root upward into the soil. Watch the action of the stem for six or eight days, and sketch it. After the stem has directed itself well up-ward, invert the pot again, and watch the growth. After a week remove the plant and notice the direction of the root. Sketch it entire, showing the changes of direction.

252, 253. — Experiment showing the direction of growth in stems: 252, young potato planted in an inverted position; 253, the same after an interval of eight days.

At the same time that this experiment is arranged, lay another pot with a rapidly growing plant on one side, and every forty-eight hours reverse the position of the pot, laying it on the opposite side. At the end of ten or twelve days remove the plant and examine. How has the growth of root and stem been affected?

What do we learn from these experiments and from those in Sections 153 and 154, as to the normal direction of growth in these two organs respectively?

160. Geotropism. — This general tendency of the growing axes of plants to take an upward and downward course — in other words to point to and from the center of the earth — is called *geotropism*. It is positive when the growing organs point downwards, as most primary roots do; negative when they point upwards, as in most primary stems; and transverse or lateral, when they extend horizontally, as is the case with most secondary roots and branches.

161. Gravity and Growth. — It has been proved by experiment that geotropism is due to gravity. It must be carefully noted, however, that the influence here alluded to is not the mere mechanical effect of gravity due to weight of parts, as when the bough of a peach or an orange tree is bent under the load of its fruit, but a certain stimulus to which the plant reacts by a spontaneous adjustment of its growing parts. In other words, geotropism is an active and not a passive function, and the plant will even overcome considerable resistance in response to it. If a sprouted bean is laid on a dish of mercury covered with a layer of water, as in Figure 254, the root will force its way

254. — Experiment showing the root of a seedling forcing its way downward through mercury.

downward into the liquid, although the mercury is fourteen times heavier than an equal bulk of the bean root substance, and the geotropism of the root must thus overcome a resistance equal to at least fourteen times its own weight.

162. Other Factors. — The direction of growth is influenced by many other factors, such as light, heat, contact

255. — A piece of a haulm of millet that has been laid horizontally, righting itself through the combined influence of contact and negative geotropism.

with other bodies, and perhaps by electricity. The result of all these forces is an endless variety in the forms and direction of organs that seems to defy all law. Heat, unless excessive, generally stimulates growth; contact sometimes simulates it, causing the stem to curve away from the disturbing object, and sometimes retards it, causing the stem to curve towards the object of contact by growing more rapidly on the opposite side, as in the stems of twining vines. Light stimulates nutrition, but generally retards growth. The heliotropic movements of plants (Sections 54–57) are effected in this way; the growth being checked on that side, the plant bends toward the light.

163. Internal Forces of the Plant. — Another important factor exists in the internal constitution of the plant itself. Place a segment of prickly pear (*Opuntia*) or other cactus, tip downward in the soil; roots will develop with great difficulty [1] because the natural forces of the plant tend to carry the root forming material to the base, and it takes time for the external factors of dampness, moisture, and gravitation to overcome this inherent tendency. Place two leafy twigs of some herbaceous plant, one in its natural position, the other bottom upwards, in a vase of water, and notice the difference in the wilting of the leaves, due to a physiological tendency in the conducting cells to carry the crude sap toward the apex.

PRACTICAL QUESTIONS

1. Why do stems of corn, wheat, rye, etc., straighten themselves after being prostrated by the wind? (162.)

2. Can a plant grow and lose weight at the same time? (155.)

3. Do plants grow most rapidly in the daytime, or at night? (162.)

4. Reconcile this with the fact that green plants will finally die if deprived of light.

5. Which would be richer in nourishment, hay cut in the evening or in the morning? Why? (24, 25, 26, 162.)

6. Which grows more rapidly, a young shoot or an old one?

7. Which, as a general thing, are the more rapid growers, annuals or perennials? Herbaceous or woody-stemmed plants?

8. Name some of the most rapid growers you know?

9. Of what advantage is this habit to them?

[1] Sachs, " Physiology of Plants."

FIELD WORK

The subjects treated in this chapter can best be studied in the laboratory, and afford little opportunity for field work, except in regard to the various adaptations for the protection and dispersal of seed. Look through the woods and fields for examples of these adaptations and explain how they are each suited to their purpose. To an imaginative mind there is something almost pathetic in what seem to be the shifts employed by the mother plants, themselves incapable of motion, to launch their offspring in the world.

Note the absence of weeds in woodlands and places remote from cultivation, and account for it. Look along railroads, along common roadsides, around wharves, factories, railroad stations, warehouses, and barnyards, for introduced plants, and account for their presence. Study the history, habits, and the local distribution of some of the common weeds of your neighborhood, and suggest means for extirpating them.

V. ROOTS AND UNDERGROUND STEMS

FUNCTION AND STRUCTURE OF ROOTS

MATERIAL. — Two earthen pots, with a growing plant in one. Some coarse netting, a common tumbler, and sprouting seeds of mustard, or other easily germinating kind. A stalk with roots, of corn or any kind of grass, and one of cotton or other woody plant. A woody taproot inserted in red ink from four to six hours before the lesson begins.·

164. Roots as Holdfasts. — One use of ordinary roots is to serve as props and stays for anchoring plants to the soil. Tall herbs and shrubs, and vegetation generally that is

256. — Dandelion : *a*, common form, grown in plains region at low altitude ; *b*, alpine form.

exposed to much stress of weather, are apt to have large, strong roots. Even plants of the same species will develop systems of very different strength according as they grow in sheltered or exposed places.

165. Root Pull. — Roots are not mere passive holdfasts, but exert an active downward pull upon the stem. Notice the rooting end of a strawberry or raspberry shoot and observe how the stem appears to be drawn into the ground at the rooting point. In the leaf rosettes of herbs growing flat on the ground or in the crevices of walls and pavements, the strong depression observable at the center is due to root pull.

257. — Raspberry stolon showing root pull.

166. Roots absorb Moisture. — Fill two pots with damp earth, put a healthy plant in one and set them side by side in the shade. After a few days examine by digging into the soil with a fork and see in which pot it has dried most. Where has the moisture gone? how did it get out?

167. Roots shun the Light. — Cover the top of a glass of water with thin netting, lay on it sprouting mustard or other convenient seed. Allow the roots to pass through the netting into the water, noting the position of root and stem. Envelop the sides of the glass in heavy wrapping paper, admitting a little ray of light through a slit in one side, and after a few days again observe the relative position of the two organs. How is each affected by the light?

168. Roots seek Air. — Remove a plant from a porous earthenware pot in which it has been growing for some time; the roots will be found spread out in contact with the walls of the pot instead of embedded in the soil at the center. Why is this?

169. Roots seek Water. — Stretch some coarse netting covered with moist batting over the top of an empty tumbler. Lay upon it some seedlings, as in Section 167, allowing the roots to pass through the meshes of the netting. (A piece of cardboard with holes in it will answer.) Keep the batting moist, but take care not to let any of the water run into the vessel. Observe the position of the roots at intervals, for twelve to twenty-four hours, then fill the glass with water to within 10 millimeters (a half inch, nearly) or less of the netting, let the batting dry, and after eight or ten hours again observe the position of the roots. What would you infer from this experiment as to the affinity of roots for water?

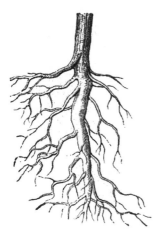

258. — Branched taproot of maple.

170. Taproots. — Gather a stalk of cotton or any hard wood shrub, and one of corn or other grain, and compare them with each other and with the roots of seedlings of the same species. Notice the difference in their mode of growth. In the first kind a single stout prolongation called a taproot proceeds from the lower end of the hypocotyl and continues the axis of growth straight downwards, unless turned aside by some external influence. A taproot may be either simple, as in the turnip, radish, dandelion, and most herbs, or branched, as in shrubs and trees generally. In this case the main axis is called the primary root, and the branches are secondary ones.

260. — Fascicled and tuberous or fusiform (secondary) roots of dahlia: *a, a,* buds on base of the stem (*after* GRAY).

259. — Fibrous root.

171. Fibrous and Fascicled Roots. — In corn and other grasses the main axis has become aborted, or failed to develop, and a number of independent branches spring from its stub, forming what are known as fibrous roots: or the base of the hypocotyl, instead of continuing downward in a single axis, may split up into a number of smaller ones.

as in the pumpkin.   When roots of this kind are thick and fleshy, they are usually described as *fascicled*.

172.   The Two Modes of Growth — This difference in the mode of growth is very apparent in the seedling, as will be evident on referring to your sketches.   The first kind is called the *axial* mode, because it is a continuation of the main axis of the plant ; the second is the *nonaxial*, or for want of a better word we may call it the *radial* mode, since the roots radiate in all directions from a common axis.

173.   Importance of this Distinction. — This distinction has important bearings in agriculture.   Roots of the first kind, which are characteristic of most dicotyledons, strike deep, and draw their nourishment from the lower strata of the soil, while the radial kind spread out near the surface and are more dependent upon external conditions.

174.   Root Structure. — Cut a cross section of any woody taproot about halfway between the tip and the ground level, examine it with a lens and sketch it.   Label the dark outer covering, *epidermis*, the soft layer just within that, *cortex*, the hard, woody axis that you find in the center, *vascular cylinder*, and the fine silvery lines that radiate from the center to the cortex, *medullary rays* (in a very young root, these will not appear).   Cut a section through a root that has stood in red ink for about three hours and note the parts colored by the fluid.

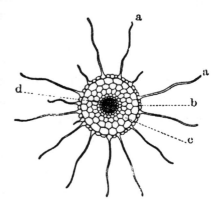

261. — Cross section of a young taproot: *a, a,* root hairs; *b,* epidermis; *c,* cortical layer; *d,* fibrovascular cylinder. Note the absence of medullary rays during the first year of growth.

What portion of the root, should you judge from this, acts as a conductor of the water absorbed from the ground ?

Make a longitudinal section through the upper half of

your specimen, continuing it an inch or two into the stem; do you find any sharp line of division between the two?

175. The Active Part of the Root. — It is only the newest and most delicate parts of the root that produce hairs and

262. — Root of a tree on the side of a gulley acting as stem.

are engaged in the active work of absorption, the older parts acting mainly as carriers. Hence, old roots lose much of their characteristic structure and take on more and more of the office of the stem, until there is practically no difference between them. On the sides of gullies, where the earth has been washed from around the trees, we often see the upper portion of the root covered with a thick bark and fulfilling every office of a true stem.

176. Use of the Epidermis. — Cut away the lower end of a taproot; seal the cut surface with wax so as to make it perfectly water-tight, and insert it in red ink for at least half the remaining length, taking care that there is no break in the epidermis. Cut an inch or two from the tip of the lower piece, or if material is abundant, from another root of the same kind, and insert it without sealing the cut surface, in red ink, beside the other. At the end of three or four hours, examine longitudinal sections of both pieces. Has the liquid been absorbed equally by both? If not, in which has it been absorbed most freely? What conclusion would you draw from this, as to the passage of liquids through the epidermis?

From this experiment we see that the epidermis, besides protecting the more delicate parts within from mechanical injury by hard substances contained in the soil, serves by its comparative imperviousness to prevent evaporation, or reabsorption by the soil, of the sap as it flows from the root hairs up to the stem and leaves.

177. The Branching of Roots. — Peel off a portion of the cortex from any branching taproot and notice the hard, woody axis that runs through the interior.   Pull off a branch from the stem and one from the root; which comes off most easily? Examine the points of attachment of the two and see why this is so.   This mode of branching from the central axis instead of from the external layers, as in the stem, is one of the most marked

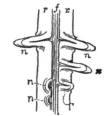

263. — Vertical section of branching root, showing the branches, *n, n*, originating in the central axis, *f*, and passing through the cortex, *r, r.*

distinctions between the structure of the two organs.

178. Distinctions between Root and Stem. — In stems the branches always occur, as we saw in our study of leaves, at regular intervals called nodes (Sec. 50), while in the root they occur quite irregularly.   The root grows only from just behind the tip; stems increase by the development of successive internodes, each of which may continue to grow for some time after the development of its successor (Secs. 153, 157).   The stem is normally an ascending, the root a descending, axis; the one bears leaves and buds at regular intervals, the other bears no leaves and only occasional buds of the kind called *adventitious;* that is, buds which appear by chance, as it were, at irregular intervals.   There are other distinctions recognized by botanists, but they are too technical to be considered here.

PRACTICAL QUESTIONS

1. Why will most plants grow so much better in an earthen pot or a wooden box than in a vessel of glass or tin?   (168.)

2. Which absorb most from the soil, plants with light roots and abundant foliage or those with heavy roots and scant foliage?

3. Which will require the deeper tillage, a bed of carrots or one of strawberries?   (173.)

4. Which will best withstand drought, a crop of cotton or one of Indian corn? Which will thrive best on high and dry ground? (173.)

5. Which will interfere least with the nourishment of the trees if planted in a peach orchard, cotton or oats? (173.)

6. Should a crop of cotton and one of hemp succeed each other on the same land?

7. Why does the gardener manure a grass plat by scattering the fertilizer on top of the ground while he digs around the roses and lilacs and deposits it underground?

8. Where should the manure be placed to benefit a tree or shrub with wide-spreading roots? (175.)

9. Is it a wise practice to mulch a tree by raking up the dead leaves and piling them around the base of the trunk, as is so often done? Why, or why not?

10. Why are willows usually selected in preference to other trees for planting along the borders of streams in order to protect the banks from washing?

FLESHY ROOTS

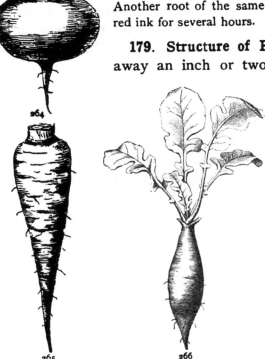

MATERIAL. — A turnip, or other fleshy root. Another root of the same kind that has stood in red ink for several hours.

179. Structure of Fleshy Roots. — Cut away an inch or two from the tip of a young fleshy root of any kind, and let it stand from six to twelve hours in red ink. Then cut into two or three equal transverse sections and observe the course of the fluid. Through what portion did it rise most readily? Sketch one of the sections and compare it with your drawing of the woody tap-

264–266. — Shapes of fleshy roots (GRAY) : 264, napiform ; 265, conical ; 266, spindle-shaped.

root. The ring of ink marks the boundary between the cortex and the central axis. Cut through one of the sections vertically and notice that the portion marked "vascular cylinder" in the hard root has here been replaced by a soft, nutritious substance. Put a drop of iodine on it and see if it contains starch. Peel off a part of the cortex and observe that the woody or conducting portion of the interior is confined principally to a thin layer on the outside of the thickened fleshy axis. Can you tell now why the course of the red ink in this kind of root is confined mainly to a ring just inside the cortex, while in hard roots — in the newer, active parts of them at least — it runs through the whole of the central axis? (Sec. 174.)

This band of woody or vascular tissue, as it is called, becomes very evident in old turnips and radishes. In the beet it is arranged peculiarly, being disposed in concentric layers alternating with the fleshy substance, instead of in a single layer next the cortex. These vascular rings give to a section of beet the appearance of certain woody stems with their rings of annual growth, but their origin is quite different.

180. Function of Fleshy Roots. — What is the use of fleshy roots? We give a practical answer to this question every time we eat a carrot or a turnip. Fleshy roots are especially useful to *biennials*, a name given to herbs that take two years to perfect their fruit, in contradistinction to *annuals*, which complete their life history in a single season. The biennials spend their first year in laying by a store of nourishment which they use up the next year in producing a crop of seed — provided man does not forestall them and appropriate it to his own use. This explains why a radish or a turnip is so dry and tasteless the second year; nearly all of its store of food has been exhausted in maturing seed.

181. Perennial Herbs are those that live on indefinitely from year to year. Many of these, like the dahlia and hawkweed, die down above ground in winter but are enabled to keep their underground parts alive through the supply of nourishment stored in their roots, and thus get the advantage of their competitors by starting out in spring with a good supply of food on hand. If you will dig around any of our hardy winter herbs, such as the rib grass (plantain), dandelion, and common dock, that keep a rosette of green leaves above ground all the year, you will generally find that they have a more or less fleshy tap-root full of nourishment, stored away underground.

PRACTICAL QUESTIONS

1. Compare a root of wild carrot with a cultivated one; what difference do you see?

2. Why are the fleshy roots of wild plants so much smaller than those of similar species in cultivation?

3. Why do farmers speak of turnips and other root crops as "heavy feeders"? (180.)

4. Which is most exhausting to the soil, a crop of beets, or one of oats? Onions, or green peas?

5. Which is best to succeed a crop of turnips on the same land, hay or carrots?

6. Write out what you think would be a good rotation for four or five successive crops.

7. Study the following rotations and give your opinion about them; suggest any improvements that may occur to you, and give a reason for the change: Beets, barley, clover, wheat; cotton, oats, peas, corn; oats, melons, turnips; cotton, oats, corn and peas mixed, melons; cotton, hay, corn, peas.

SUB-AËRIAL ROOTS

MATERIAL. — A hyacinth bulb or a cutting of wandering Jew grown in a glass of water. Specimens of any kind of parasitic plants that can be obtained, such as mistletoe, dodder, resurrection fern (*Polypodium incanum*), etc. Freshly rooted cuttings of geranium, coleus, or other easily rooting twig.

182. Subterranean and Sub-aërial Roots. — The roots we have been considering are all subterranean and bring the plant into relation with the earth, whether for purposes of nourishment, or of anchorage to a fixed support, or, as in the majority of cases, for both. But many plants do not get their nourishment directly from the soil, and these give rise to the various forms of sub-aërial roots, or those that grow above ground.

183. Water Roots. — Large numbers of plants are adapted to live in the water, either floating freely, as the duckweed (*Lemna*) and bladderwort (*Utricularia*), or anchored to mud and sticks on the bottom. Water roots are generally white and threadlike and more tender and succulent than ordinary soil roots. Many land plants will develop water roots and thrive on that element if brought into contact with it. Place a cutting of wandering Jew in a clear glass of water, and in from four to six days it will develop beautiful water roots in which both hairs and cap are clearly visible to the naked eye.

The chief office of ordinary roots being to absorb moisture, they have a great affinity for water, and its presence or absence exerts a strong determining influence on their direction, often overcoming that of geotropism (Sec. 169).

184. Parasitic Plants are those that live by attaching themselves to some other living organism, from which they draw their nourishment ready made. Their roots are adapted to penetrating the substance of the *host*, as their

267, 268. — Mistletoe penetrating bough of oak : 267, lower part of stem attached to branch ; 268, longitudinal section through one of the haustoria strands, showing its progress as the branch thickens.

victim is called, and absorbing the sap from it. They are appropriately named *haustoria*, a word meaning suckers, or absorbers. Dodder and mistletoe are the best-known examples of plant parasites, though the latter is only partially parasitic, as it merely takes up the crude sap from the host and manufactures it into food by means of its own green leaves.

185. Saprophytes are plants like the Indian pipes (*Mono-tropa*) and squaw root (*Conopholis*) that live upon dead and decaying vegetable matter. They are only partially par-

asitic, and do not bear the haustoria of true parasites. A good many plants that appear to live an honest life above ground practice a secret parasitism by sending their roots into

269. — Roots of Gerardia parasitic underground (*after* GRAY).

those of their neighbors beneath the soil and drawing part of their nourishment from them. Among those that show a propensity to this degenerate habit are the pretty yellow gerardias, and their kindred, the yellow rattle (*Rhinanthus*), and the Canada lousewort (*Pedicularis*).

186. Aërial Roots are such as have no connection at all with the soil or with any host plant, except as they may lodge upon the trunks and branches of trees for a support. In our climate aërial roots are generally subsidiary to soil roots, like the long dangling cords that hang from some species of old grape vines ; or they subserve other purposes altogether than absorbing nourish-

270. — A small orchid with aërial roots, growing on the bough of a tree (*after* GRAY).

ment, as the climbing roots of the trumpet vine and poison ivy.

187. Adventitious Roots is a name applied to any kind that occur on the stems of plants or in other unusual positions. Common examples are the roots that put out from the lower nodes of corn and sugar cane and serve both to supply additional moisture and to anchor the plant more firmly to the soil. Most plants will develop adventitious roots if covered with earth or even if merely kept in contact with the ground. The gardener takes advantage of this property when he propagates by cuttings or layers.

Place a cutting of rose geranium or of coleus in a pot of moist sand. As soon as the roots begin to form, examine the stem with a lens to see from what portion they spring

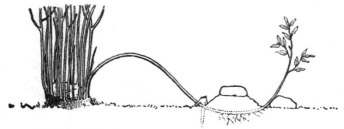

271. — New stocks with adventitious roots produced by layering.

— whether from near the circumference, or from the center. What part of the stem should you infer from this, is most actively concerned in the work of growth?

PRACTICAL QUESTIONS

1. Do the adventitious roots of such climbers as ivy and trumpet vine draw any nourishment from the objects to which they cling?

2. How do you know this?

3. Do they injure trees by climbing upon them; and if so, how?

4. What is the use of the aërial roots of the scuppernong grape?

5. Is the resurrection fern (*Polypodium incanum*) a parasite or an air plant?

6. On what plants in your neighborhood does mistletoe grow most abundantly? Dodder?

7. Is mistletoe injurious to the host?

8. Name some plants that are propagated mainly, or solely, by roots and cuttings.

UNDERGROUND STEMS

MATERIAL. — Underground stems of couch grass, nut grass, violet, iris, or any rootstocks obtainable. In cities, if nothing better is to be had, some dried orris root or calamus might be obtained from a druggist. Any kind of tuber, such as potato, artichoke, Madeira vine, etc. A sweet potato. A scaly lily bulb and one of onion or hyacinth. Potatoes and sweet potatoes treated with red ink.

188. Rootstocks. — So like fleshy roots are certain thickened underground stems that it is not always easy to distinguish between them. So long as the stem remains above ground there is little danger of mistaking its identity, even when it puts forth roots from every node, like the creeping stems of Bermuda grass and couch grass. Even in such underground stems as those of the mint and couch grass their real nature is evident from

273. — Rootstock of creeping panic grass.

272. — Running rootstock of peppermint (GRAY).

the regular nodes into which they are divided, and the scales which they bear instead of leaves. Stems of this kind are called rootstocks. They usually send out roots from every node and are the most ineradicable pests the farmer has to contend with, since each joint is capable of developing into a new plant, and chopping them to pieces serves only to aid in their propagation.

189. Rhizomas. — Rootstocks do not always retain their stemlike nature so plainly, but are commonly more or less shortened and thickened, as in the violet, iris, bulrush, sweet flag, bloodroot, etc., and it is to this condition that the name rhizoma is usually applied. A typical example of

274. — Rhizoma of Solomon's seal (*after* GRAY).

the rhizoma is that of the Solomon's seal (Fig. 274.) The peculiar scars from which it takes its name are caused by the falling away each year of the flowering stem of the season, after its work is done, leaving behind the joint or node of the underground stem from which it originated. Thus the plant lives on indefinitely, growing and increasing at one end as fast as it dies at the other. The joints on the rhizoma mark, not the age of the plant, but of each joint or internode. If there are two or three joints, this indicates that the oldest of them is two or three years old, as the case may be.

Examine a rhizoma of the iris, or any other specimen obtainable. How many joints do you find? Where is the oldest? How old is it? Are they all entirely underground? Where do the true roots spring from? The flower stems? Notice the rings or ridges that run across the upper side of each joint. These are the leaf scars, and each scar marks one of the very short internodes of the past season's growth. At the nodes, in the axils of the leaf scars, buds frequently occur, producing other joints, which may be considered branches, and it is these branches that give to the rootstocks of the iris and blackberry lily their thick, matted appearance. How many leaves did last year's joint of your specimen bear, and how many internodes had it?

190. Tubers. — When a rhizoma is very greatly enlarged, as in the artichoke and potato, it is called a *tuber*. Its real nature in such cases is often very much disguised, but a little study will make it clear. The so-called root of wild smilax shows very plainly the gradations from leaves to scales and from stem to tuber.

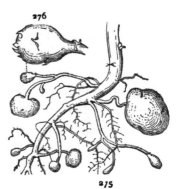

In the typical tuber, of which the potato is the most familiar example, the internodes are so thickened and shortened as to have lost all resemblance to a stem, but their nature is revealed by the *eyes*. These are really nothing else than buds growing in the axils of leaves, which are represented in the potato by the

275, 276. — Tubers (*after* GRAY): 275, forming potatoes; 276, young potato enlarged.

little scale that forms the lid to the eye. (In an old potato the scales will probably have disappeared; try to get fresh ones for examination, and if possible, with some of the attaching stems still remaining. The artichoke and tubers of the Madeira vine also make good objects for study.) Notice the arrangement of the eyes, or buds; is it alternate or opposite? How many ranked? Make a sketch of the potato as it appears on the outside. Make a similar sketch of the sweet potato, and compare the two. Is there any scale below the eye in the sweet potato? Do the eyes occur in any regular order?

191. Make a cross section of each, and sketch them. Notice the thin dark ring that runs around the inside of the potato at some distance from the circumference. Label this *vascular tissue;* the loose porous layer between it and the skin, *cortex;* the central portion within the vascular ring, *pith;* and the outer skin, *epidermis.* See if you can find corresponding parts in the sweet potato, and label them.

Put one of the cut ends of each in red ink (this should have been attended to before the recitation), let them stand four to five hours, then make sections parallel to the cut surface till you reach the point where the red ink has penetrated; what difference do you notice? Which has the thicker cortex? Compare the behavior of the potato with that of the turnip treated with red ink in Section 179. What would you infer from this as to the office of the woody tissue? What is the office of the epidermis? If you are in doubt, peel a tuber and weigh it. At the same time weigh one of about the same size from which the skin has not been removed, and put the two side by side in a dry place. At the end of three or four days weigh them again and see which has lost the most.

We have learned that roots are not divided into nodes, that they never bear leaves, that they branch quite irregularly, and that they sometimes bear adventitious buds. Now can you state some of the reasons why the potato is regarded as a stem and the sweet potato as a root?

192. Storage of Nourishment. — The object of both is the same, the storage of nourishment. Drop a little iodine on each and see what this nourishment consists of. Which contains the more starch?

It is this abundant store of food that makes the potato such a valuable crop in cold countries like Norway and Iceland, where the seasons are too short to admit of the slow process of developing the plant from the seed.

193. The Bulb is a form of underground stem reduced to a single bud. Get the scaly bulb of a white garden lily, and sketch it from the outside and in cross and vertical section. Compare it with the scaly winter buds of the oak and hickory or other common deciduous tree. Make an enlarged sketch of the latter on the same scale as the lily, and the resemblance will at once become clear. The scales of the bulb are, in fact, only thick, fleshy leaves

277.— Scaly bud of oak enlarged.

278. — Scaly bulb of lily (GRAY).

279. — Bulblets in the axils of the leaves of a tiger lily (GRAY).

closely packed round a short axis that has become dilated into a flat disk. From the terminal node of this transformed stem, *i.e.*, the center of the disk, rises the flower stalk, or *scape*, as it is called, of the season. After blossoming, the scape perishes with its bulb, and their place is taken by new ones which are developed from the axils of the scales, thus revealing their leaflike nature.

That bulbs are only modified buds is further shown by the bulblets that sometimes appear among the flowers of the onion, and in the leaf axils of certain lilies. They never develop into branches, but drop off and grow into new plants just as the subterranean bulbs do.

194. Tunicated Bulbs. — Compare an onion or a hyacinth bulb with a lily bulb. In what respect does it differ from the lily bulb? Pull off the outer layers, which have become dry and papery, and observe how the inner fleshy

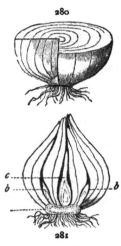

ones encircle one another successively. A bulb of this kind, made up of successive layers, is said to be *tunicated.* Look for the flower cluster in the center. Do you find any axillary bulbs? Any axillary scapes? Compare the concentric rings of a tunicated bulb as seen in cross section with those of the beet; are they of the same nature? Before answering, look again at your cross section of the lily bulb and think what would happen if the scales were to be broadened sufficiently at the base for each one to encircle completely all within it. Compare the leaves and scales of

280, 281. — Tunicated bulbs (GRAY): 280, cross section of an onion; 281, vertical section of a tulip bulb, showing terminal bud, *c*, and axillary buds, *b, b.*

the onion with the leafstalk of the sycamore (Fig. 20), and see if you can find any reason for regarding them as modified petioles.

195. Uses of Underground Stems. — Though the chief function of underground stems is the storage of nourishment, they serve other purposes also. In plants like the ferns, that require a great deal of moisture, and in

282. — Leaf of an onion divided lengthwise.

others growing in dry places, like the blackberry lily, that need to husband it carefully, they may be useful in preventing the too rapid evaporation that would take place through aërial stems. Defense against frost, cold, heat, and other dangers, as well as quickness of propagation, are also attained, or assisted by this means.

PRACTICAL QUESTIONS

1. Name some plants in your neighborhood that are propagated by rootstocks; by rhizomas; by tubers.

2. What is the advantage of propagating in this way over planting the seed? (192, 195.)

3. What other advantages, if any, does each of the plants named gain from its earth-seeking (geophilous) habit?

4. What makes the nut grass so troublesome to farmers?

5. Is its nut a root, or a tuber? How can you tell? (190, 191).

6. Suggest some ways for destroying weeds that are propagated in this way.

7. Could you get rid of wild onions in a pasture by mowing them down? By digging them up? (193.)

8. Is it wise for farmers to neglect the appearance of such a weed in their neighborhood, even though it does not infest their own land?

9. Name any plants of your neighborhood, either wild or cultivated, that are valued for their rhizomas; for their tubers.

10. What part of the plants named below do we use for food or other purposes? Ginger, angelica, ginseng, cassava, arrowroot, garlic, onion, sweet flag, iris, sweet potato, Cuba yam, artichoke.

11. Why are the true roots of bulbous and rhizome bearing plants generally so much smaller in proportion to the other parts than those of ordinary plants? (192, 195.)

12. If the Canada thistle grows in your vicinity, examine the roots and see if there is anything about them that will help to account for its hardihood and persistency.

13. If you live in the region of the horse nettle (*Solanum carolinense*), explain how it is enabled to flourish in such hard and forbidding places.

PLANT FOOD

MATERIAL. — An ounce or two each of different kinds of seeds, and a lamp stove or other convenient means of drying them. A pair of scales.

196. Solids, Liquids, and Gases. — The habit of storing up food in some part of their structure for future use is practically universal among plants. Let us now inquire what this food consists of and where it comes from.

Take a quantity of seeds of different kinds (about thirty grams of each, or one ounce approximately, will answer), weigh each kind separately and then dry them at a high temperature, but not high enough to scorch or burn them. After they have become perfectly dry, weigh them again. What proportion of the different seeds was water, as indicated by their loss of weight in drying?

Burn all the solid part that remains, and then weigh the ash. What proportion of each kind of seed was of incombustible material? What proportion of the solid material was destroyed by combustion?

Test in the same way the fresh, active parts of any kind of ordinary land plant (sunflower, hollyhock, pea vines, etc., make good specimens) and of some kind of succulent water or marsh plant (Sagittaria, water lily, fern, etc.). Do you notice any difference in the amount of water given off and of solid matter left behind? In the character of the ashes left? Have you observed in general any difference between the ashes of different woods; as, for instance, hickory, pine, oak, etc.?

197. Essential Constituents. — The composition of the ash of any particular plant will depend upon two things: the absorbent capacity of the plant itself and the nature of the substances contained in the soil in which it grows. But chemical analysis has shown that however the ashes may vary, they always contain some proportion of the following substances: potassium (potash), calcium (lime), magnesium, phosphorus, and (in green plants) iron. These elements occur in all plants, and if any one of them is absent, growth becomes abnormal if not impossible.

283. — Water cultures of buckwheat, showing effect of the lack of the different food elements: 1, with all the elements; 2, without potassium; 3, with soda instead of potash; 4, without calcium; 5, without nitrates or ammonia salts.

The part of the dried substances that was burned away after expelling the water consists, in all plants, mainly of carbon, hydrogen, oxygen, nitrogen, and sulphur, in varying proportions. These five rank first in importance among the essential elements of vegetable life, and without them the plant cell itself, the physiological unit of vegetable structure, could not exist. They compose the greater part of the substance of every plant, carbon alone usually forming about one half the dry weight.   Other substances may be present in varying proportions, but the two groups named above are found in all plants without exception, and so we may conclude that (with the possible addition of chlorine) they form the indispensable elements of plant food.   Carbon, hydrogen, oxygen, nitrogen, sulphur, and phosphorus compose the structure of which the plant is built.   The other four do not enter into the substance as component parts, but aid in the chemical processes by which the life functions of the plant are carried on, and are none the less essential elements of its food.   Figure 283 shows the difference between a plant grown in a solution where all the food elements are present and others in which some of them are lacking.

198. How Plants obtain their Food. — With a few doubtful exceptions, plants cannot assimilate their food unless it is in a liquid or gaseous form.   Of the gases, carbon dioxide, oxygen, and hydrogen can be freely absorbed from the air or from water with various substances in solution, but most plants are so constituted that they cannot absorb free nitrogen from the air; they can take it only in the form of compounds from nitrates dissolved in the soil, and hence the importance of ammonia and other nitrogenous

284. — Roots of soy bean bearing tubercle-forming bacteria.

compounds in artificial fertilizers.   Some of the pea family, however, bear on their roots little tubers containing minute organisms called bacteria, which have the power of extracting nitrogen directly from the free air mingled with the soil; and hence, wherever these tuber-bearing legumes are present the soil is found to be enriched with nitrogen in a form ready for use.

Plants also obtain their supply of the various **mineral salts** needed by them from solutions in the soil water which they absorb through their roots. Different species, and even different varieties of the same species, absorb these substances in very different proportions, and upon this fact, much more than upon the form of roots (Sec. 173), depends the principle of the rotation of crops in farming.

199. Plants can not choose their Food. — Substances are often found in plants which appear to be useless, and some, as zinc and lead, which are positively harmful. This shows that the roots are not able to choose their own nourishment, but absorb whatever is present in the soil in soluble form and can penetrate their cell walls. It is not safe, therefore, to conclude merely because a substance is found in a plant that it should constitute a part of its food. Neither can we always be sure that because a plant will grow in a certain soil this is the best soil for it, since its presence there may be due merely to adaptation or toleration, and it might do better if given a chance somewhere else. All these circumstances present matter for careful discrimination by the farmer.

200. Food Manufacture. — The proportion of ash found in green plants increases from the roots upwards to the leaves, thus showing that the latter are the organs in which the manufacturing or building-up process takes place, and its products are most abundant there. The first article of food to be recognized is starch, but others also occur and are distributed to the parts where they are needed. Of course solid substances like starch and the various ashes that we find in the structure of plants can not pass through the walls of the cells unchanged, but must be reduced to the form of a solution. In the case of substances that are insoluble they must first be transformed to soluble ones and then reformed into their original constitution, so we see that the nutrition of plants is a very complicated process, involving repeated chemical changes and redistributions of material.

1. Will a pound of pop corn weigh the same after it has been "popped"? (196.)

2. Could any plant grow in a soil from which nitrogen was entirely lacking? Phosphorus? Potash? Lime? (197.)

3. Could it live in an atmosphere devoid of oxygen? Nitrogen? Carbon dioxide? (197.)

4. Is the same kind of fertilizer equally good for all kinds of soil? For all kinds of plants? (198.)

5. Is starch soluble in water?

6. How does it get from the leaves where it is manufactured to the rootstocks where it is stored? (200.)

7. Why does too much watering interfere with the nourishment of plants?

8. Are ashes fit for fertilizers after being leached for lye? (197, 198.)

9. Why will any but very small shrubs be dwarfed, or make very slow growth in pots? (197, 198.)

Examine the underground parts of hardy winter herbs in your neighborhood, and of any weeds or grasses that are particularly troublesome, and see if there is anything about the structure of these parts to account for their persistence. Note the difference in the roots of the same species in low, moist places and in dry ones; between those of the same kind of plants in different soils; in sheltered and in exposed situations. Study the direction and position of the roots of trees and shrubs with reference to any stream or body of water in the neighborhood. (The elm, fig, mulberry, and willow are good subjects for such observations.) Notice also whether there is any relation between the underground parts and the leaf systems of plants in reference to drainage and transpiration.

Observe the effect of root pull upon low herbs. Look along washes and gullies for roots doing the office of stems, and note any changes of structure consequent thereon. Study the relative length and strength of the root systems of different plants, with reference to their value as soil binders, or their hurtfulness in damaging the walls of cellars, wells, sewers, etc. Dig your trowel a few inches into the soil of any grove or copse you happen to visit, and note the inextricable tangle of roots, and consider the fierce competition for living room in the vegetable world that it implies.

Tests might be made of the different soils in the neighborhood of the schoolhouse by planting seeds of different kinds and noting the rate of germination; first, without fertilizers, then by adding the different elements in succession to see which is lacking. The field for study suggested by this subject is almost inexhaustible.

VI. THE STEM PROPER

STEM FORMS AND USES

MATERIAL. — Stems of various kinds — woody, herbaceous, round, square, triangular, jointed, upright, etc. Herbaceous stems are not abundant in winter, but a few hardy herbs like shepherd's purse, dandelion, winter cress (*Barbarea*), dead nettle, or some of the garden biennials can generally be found. Of triangular and jointed stems any of the sedges and grasses will furnish examples. Have young specimens of some kind of twining stems raised in the schoolroom. Hop and morning-glory make very good examples.

201. Woody and Herbaceous Stems. — Aërial stems, or those above ground, are commonly ranked in two general classes, *woody* and *herbaceous*. The latter are more or less succulent, and die down after fruiting; the former live on from year to year, sometimes, as in the case of the giant sequoias of California and some of the primitive cypresses of our own southern swamps, even for thousands of years. Many herbaceous stems, like the garden geraniums and the common St. John's-wort, show a tendency to become woody, especially toward the base, and live on from year to year. Woody-stemmed annuals, like the cotton and castor-oil plant are not, properly speaking, herbs. In the tropical countries to which they belong, they are perennial shrubs, or even small trees, but on being transplanted to colder regions, have taken on the annual habit as an adaptation to climate.

202. Direction and Habit of Growth. — As to manner of growth, there are all forms, from the upright boles of the beech and pine to the trailing, prostrate, and creeping stems of which we have examples in the running periwinkle, the prostrate spurge, and the creeping partridge berry (*Mitch-*

ella), respectively. Prostrate and trailing stems are very apt to become creepers by the development of adventitious roots at their nodes, wherever they come in contact with the soil. Between the extremes of prostrate and upright, stems may be in various degrees,

Assurgent, that is, ascending, like common crab grass and spotted spurge (*Euphorbia maculata*), or,

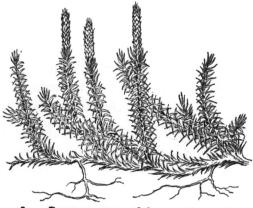

285. — Prostrate stem of Lycopodium with assurgent branches.

Declined and drooping, as the garden jessamine, the matrimony vine (*Lycium*), and some of the garden spireas. The most interesting of all in their mode of growth are the various forms of

203. Twining and Climbing Stems. — The former rise from the ground by twisting themselves spirally round their support, like the morning-glory, hop, and yellow jessamine; the latter by attaching themselves to other objects by means of adventitious roots and tendrils, as the Virginia creeper, poison ivy, pea, grape, smilax, etc. A curious fact about twiners is that with one or two exceptions each species always coils uniformly in the same direction and can not be made to change. Raise a young hop or morning-glory plant in the schoolroom, notice whether it starts to coil from right to left or from left to right, and see if you can coax it to grow in the opposite

286   287

286, 287. — Twining stems: 286, hop twining with the sun; 287, convolvulus twining against the sun.

direction.   When it has reached the end of its stake suffer it to grow about five centimeters (two inches, approximately) beyond, and watch the revolution of the tip.   Cut a hole through the center of a piece of cardboard about fourteen centimeters (five to six inches) in diameter, slip it over the loose end of the stem, and fasten it to the stake in a horizontal position with a pin.   Note the position of the stem tip every two hours and mark on the cardboard; how long does it take to complete a revolution?

204. The Cause of Twining is believed to be unequal growth on the two sides of the stem (Sec. 162) which causes the tip to revolve slowly in a spiral toward the side where growth is slowest.   Run a gathering thread in one side of a narrow strip of muslin, about a meter (one yard, approximately) long, and notice how the ruffle thus drawn will curl into a spiral when allowed to dangle from the needle.   In the same way the tension resulting from unequal growth causes the stems and tendrils of climbing plants to form themselves into spirals.

Hardly any kind of stem grows at a uniform rate in all its parts.   Ordinarily the inner part grows most rapidly. Split the stem of a fresh dandelion, hyacinth, or other herbaceous scape longitudinally, and immerse it in fresh water for 30 to 45 minutes.   Notice how the two halves curve outward, or even coil up like a watch spring.   This is on account of the tension caused by the more rapid absorption of the internal tissues, which, when relieved of the resistance of the outer wall, or epidermis, stretch themselves, as it were, but are held back and drawn into a curve by the resistance of the slower growing outer parts, as the muslin of our ruffle was curled by the gathering thread.

205. The Object of the Various Forms of Stem Growth is in all cases the same; to bring the leaves into the best possible relations with the light and air.   The stem, besides other important uses, serves as a mechanical support, or framework, to bind the other organs together, and they are largely dependent upon it for proper exposure to light and air. In general, leaves seek the best possible light exposure, and hence the normal growth of the stem is upward, toward the light. There are exceptions, however, in the case of shade-loving plants that seek the shelter of the forests, and certain winter green herbs like the chickweeds, Indian strawberry, and dandelion, that protect themselves against stress of weather by lying low and hugging the earth. The same habit may temper both the summer's heat and winter's cold, by shading the earth around the roots and preventing too rapid evaporation in the hot season, and by keeping them in contact with the warm earth and preventing too rapid radiation in winter.

206. The Surface of Stems, like that of leaves, may be hairy, prickly, smooth, rough, etc., and the same terms are used in describing them. The object of these adaptations is the same as in leaves. Grooves and wings and hairs may either be related to drainage and aid in the conduction of water, or they may help or hinder the visits of certain insects and other animals. Some of these devices are very ingenious, and have been imitated by man. The sticky gum exuded from the upper nodes of the catchfly (*Silene*) protects the flower against the visits of crawling insects as effectively as would a strip of sticky fly paper; and our barbed-wire fences do not serve their purpose any better than the prickles of the black-berry and the cactus. In regard to

207. Shape, stems are either round (terete), flattened, square, triangular, etc. Sometimes the shape is of great use in helping to distinguish different kinds of plants. In the mint family and its allies, square stems are prevalent; the sedges are generally characterized by triangular ones, and grasses by round, hollow, jointed *culms*, or *haulms*, as they are called, like those of wheat, oats, and rye.

288.—Culm of millet.

208. Runners and Stolons, of which we have familiar examples in the strawberry and currant respectively, are

stems or branches by which plants propagate themselves above ground as readily as by rootstocks underground. Suckers are shoots from adventitious root buds. The rose, raspberry, blackberry, and asparagus are propagated almost entirely by their means. The little shoots, called by gardeners scions, that spring up around the foot of apple and pear trees, and many others, have a similar origin.

289. — Orange hawkweed with runners.

209. Modifications of the Stem. — Like leaves, the stem is subject to many modifications, and is made to serve various purposes other than its normal ones. With some of these we have already become acquainted in its underground condition. Aërial stems frequently serve like purposes. The sugar cane carries a rich supply of sweets in its juicy internodes, and cabbage stalks also are well stocked with food before flowering. In the cactus family, which inhabit dry and desert regions, where the scanty moisture they draw from the earth would be too rapidly exhaled from the expanded surface of leaves, the foliage has either disappeared altogether or been reduced to mere spines, while the greatly thickened stems have taken upon themselves the triple office of leaf, stalk, and storeroom. Examine a potted cactus, or a

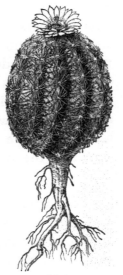

290. — Melon cactus, showing greatly condensed stem for the storage and preservation of moisture.

joint of the common prickly pear, and notice how the whole plant has been compacted into a form that exposes the least possible extent of surface in proportion to the substance contained in it.

210. Weapons of Defense. — Examples of these may be seen in the thorns of the honey locust, the hawthorn, and old field plums. An examination of the haw, crab tree, plum, and pear will show stems in all stages of transformation from short, stubby branches to well-defined thorns. This kind of thorn must not be confounded with briers or prickles like those of the rose and smilax, which are mere appendages of the epidermis, while thorn branches

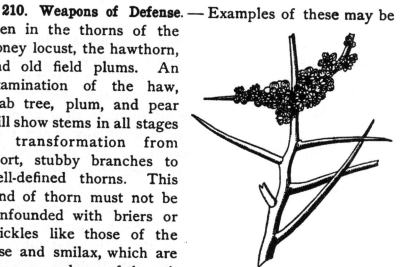

291. — Thorn branches of *Holocantha Emoryi*, a plant growing in arid regions.

have their origin in the wood beneath. They usually come from adventitious buds.

211. Stems as Tendrils. — Stems are also frequently

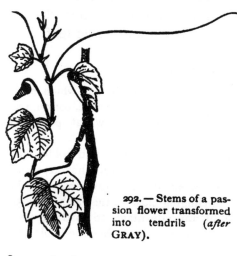

292. — Stems of a passion flower transformed into tendrils (*after* GRAY).

met with under the form of tendrils. As normal buds and branches never grow except from the axils of leaves, this kind of tendril can always be recognized by its position. In the grape and Virginia creeper, where they appear opposite the leaves on alternate sides of the stem, they represent terminal flower buds which have been pushed aside by stronger lateral ones (Sec. 245).[1] The usurping bud continues the growth of the shoot until it is in turn displaced by some succeeding lateral one, and so on, forming a succession of apparently lateral tendrils.

212. Stems as Foliage. — When branches take the place of foliage, as they not infrequently do, they are generally

so much disguised that it is difficult to recognize them, but a little attention to their point of origin will usually make their nature clear. The asparagus has already been referred to (Sec. 68). Still more striking examples are found in the butcher's broom of Europe (*Ruscus aculeatus*) and the pretty little Myrsiphyllum of the greenhouses, wrongly called smilax, that is so much used for decoration. The green blades of these plants, which are commonly regarded as foliage, are not true leaves, but curiously shortened and flattened branches that have taken upon themselves the office of leaves. Their real nature is shown by the fact that they spring each from the axil of a little scale or bract that represents the true leaf.

293. — Stem leaves (cladophylls) of a ruscus, bearing flowers.

PRACTICAL QUESTIONS

1. Which of the stems named below are woody, and which herbaceous, or suffrutescent? Blackberry, hollyhock, pokeweed, cotton, okra, morning-glory, asparagus, garden sage, reed, corn, wheat, periwinkle, sunflower, strawberry, bear's grass, broom straw.

2. Why is it that so many, both of hot-weather and cold-weather herbs, for example, knotweed (*Polygonum aviculare*), purslane, spurge, carpet weed (*Mollugo*), winter chickweed, Indian strawberry, and dandelion, all adopt the same habit of clinging close to the earth? (205.)

3. Would such a habit be of any advantage to roadside weeds and other herbs growing in exposed places where they are liable to be trodden upon and bitten by cattle?

[1] See also Gray's " Structural Botany," page 54, § 110.

4. Is there any difference in the height of the stem of a dandelion flower and a dandelion ball?

5. Of what advantage is this to the plant?

6. By what means does the gourd climb? the butter bean? the English pea? trumpet honeysuckle? grape? maypop? smilax? Virginia creeper? clematis?

7. Why do we "stick" peas with brush, and hops with poles?

8. Are gourds, watermelons, squashes, pumpkins, etc., naturally climbing, or prostrate?

9. Why does not the gardener provide them with poles or trellises to climb on?

10. Name some plants the stems of which are used as food.

11. Name some stems from which useful articles, such as sugar, gums, and medicines are obtained.

12. Do twining plants grow equally well on horizontal and upright supports? (159, 160, 244.)

13. If there is any difference, which do they seem to prefer?

STEMS OF MONOCOTYLEDONS

MATERIAL. — A stem of smilax, asparagus, or other monocotyledon that has stood in red ink for three to six hours. A dried cornstalk; the handle of a palm-leaf fan. (It would be better, of course, to have all specimens fresh, if possible, and for those who live in the southern States fresh stalks of sugar cane, palmetto, or yucca, will afford admirable objects for study.)

213. Examination of a Monocotyledonous Stem. — Take one of the dry cornstalks that can be found in the fields, almost anywhere, and study its external characters. How are the internodes divided from one another? What is the use of the very firm, smooth epidermis? Notice a hollow, grooved channel running down one side of the joints, or internodes; does it occur in all of them? Is it on the same side or on opposite sides of the alternate internodes? Follow one of these grooves to the node from which it originates; what do you find there? (In a dried stalk the bud will probably have disappeared, but traces of it can usually be found.) After studying the internal

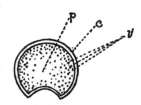

294. — Cross section of a stalk of corn: *v*, fibrovascular bundles; *c*, cortex; *p*, pith.

structure of the stalk you will understand why this groove should occur on the side of an internode bearing a bud or fruit.

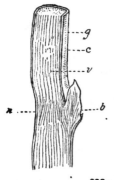

Cut a cross section midway between two nodes, and observe the composition of the interior; of what does the bulk of it appear to consist? Notice the arrangement of the little dots like the ends of cut-off threads that are scattered through the pith; where do they appear to be most abundant, toward the center or the circumference?

295. — Vertical section of cornstalk: g, groove; c, cortex; v, fibrovascular bundles mingled with parenchyma; b, bud; n, node.

Make a vertical section through one of the nodes. Cut a thin slice of the pith, hold it up to the light, and examine it with a hand lens. Observe that it is composed of a number of tiny oblong compartments or cells packed together like bricks in a wall. These are dry and empty now, but in the living stem were filled with nourishing fluids consisting of *protoplasm* and cell sap (Sec. 9), and formed what is known to botanists as the *parenchyma*, a word meaning parent tissue, because from it all the other tissues are derived.

296. — Vertical section of a portion of the interior of a dry cornstalk as seen under the lens, showing the cellular structure of the parenchyma: v, fibrovascular bundles; p, pith, or parenchyma.

Draw out one of the woody threads running through the pith. Break away a bit of the epidermis and see how very closely they are packed on its inner surface. Trace the course of the veins in the bases of the leaves that may be found clinging to some of the nodes; find their point of union with the stem; with what part of it do they appear to be continuous? Has this anything to do with the greater abundance of fibers near the epidermis? Can you follow the fibers through the nodes, or do they become confused and intermixed with other threads there? (If sugar cane is used for this study, the ring of scars left by the vascular bundles as they pass from the leaves into the stem will be seen beautifully marked just above the nodes.)

If there is an eye or bud at the node, look and see if any of the threads go into it. Can you account now for the depression that occurs in the internode above the eye or bud?

Make drawings of both cross and vertical sections showing the points brought out in your examination of the cornstalk.

214. The Vascular System. — To find out the use of the threads that you have been tracing, examine a piece of a living stem of wild smilax or other monocotyledon that has stood in red ink for three to twenty-four hours. (If the specimen stands in the coloring fluid too long the dye will gradually percolate through all parts of it. If this should be the case, look for the lines that show the ink most plainly.) Notice the course the coloring fluid has taken; what would you infer from this as to the office of the woody fibers?

These threads constitute what is called the *vascular system* of the stem, because they are made up, to a large extent, of little *vessels* or *ducts*, along which the sap is conveyed from the roots to the leaves and back from the leaves to the root and stem after it has been elaborated into food. They are, so to speak, the water pipes that supply the leaf community with the liquid nourishment which it works up into food during the process of photosynthesis (Sec. 24).

215. The Stem as a Water Carrier. — We see from this, that the stem, besides serving as a mechanical support, is the natural line of communication between the roots, where the raw material for feeding the plant is gathered, and the leaves, where this material is manufactured into food. After the sap is there elaborated and the surplus moisture given off by transpiration, the nourishment is

returned to be distributed to the other organs. Even the
roots can not be fed by the liquid they absorb from the soil
until it has been elaborated in the leaves, just as our
bodies can not be sustained by what we eat and drink until
·it has been digested in our stomachs. Hence, if the leaves
of a tree are diseased or destroyed by ignorant pruning,
the roots will suffer and die just as the leaves do if the
roots are injured.

On account of this double line of communication which
they have to maintain, the vascular threads, or *bundles*, as
they are technically called, are double; one set, composed
of larger ducts, carrying water up, and another set of
smaller ones bringing back the digested food. Can you
give a reason for their difference in size?

216. Woody Monocotyledons. — Examine sections of
yucca, smilax, or of pálmetto from the handle of a fan,
and compare them with your sketches of the cornstalk.
In which are the vascular fibers most abundant? Which
is the toughest and strongest? Why? Trace the course
of the leaf fibers from the point of insertion to the
interior. How does it differ from that
of the fibers in a cornstalk?

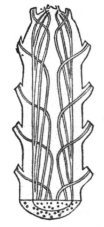

297. — Longitudinal
section through the
stem of a palm, show-
ing the curved course
of the fibrovascular
bundles (GRAY, *after*
FALKENBERG).

**217. Growth of Monocotyledonous
Stems.** — Refer to the experiment in Section 43 ; refer also to what has just been
learned regarding the course of the leaf
veins at the nodes of the cornstalk (Sec.
213), and you will have no difficulty in
identifying these veins as part of the vascular system. Each successive leaf sends
its vascular bundles down into the main
system of the stem, and any increase
in the diameter of monocotyledons takes
place by the intercalation of new bundles
from the leaves as they develop at the
nodes above. In jointed stems like the
corn and sugar cane and other **grasses,**

this intercalation takes place, as we have seen (Sec. 213), at the nodes, forming the hard rings known as joints, but in other monocotyledons the fibers entering the stem from the leaves generally tend first downwards, towards the interior (Fig. 297), then bend outward toward the surface, where they become entwined with others and form the tough, inseparable cortex that gives to palmetto and bamboo stems their great strength.

This addition of fresh vascular bundles as the axis lengthens will explain why the lower joints of cornstalks and sugar cane are so much more hard and woody than the upper ones. Generally, however, monocotyledonous stems do not increase in diameter after a certain point, and as they can contain only a limited number of vascular fibers, they are incapable of supporting an extended system of leaves and branches. Hence this class of plants, with a few exceptions, like smilax and asparagus, are

298.—A palm tree, showing the tall, branchless trunk of monocotyledons.

characterized by simple, columnar stems, and a limited spread of leaves. The cabbage palmetto, banana, and Spanish bayonet (*Yucca aloifolia*) are familiar examples in the warmer parts of our country.

218. Strength of the Monocotyledonous Structure. — Stems of this class are admirably adapted by their structure to the purposes of mechanical support. It is a well-known law of mechanics that a hollow cylinder is a great deal stronger than the same mass would be in solid form, as may easily be tested by the simple experiment of breaking in your fingers a cedar pencil and a joint of cane or a stem of smilax of the same weight. In stems that may be technically classed as solid in structure, like the corn and palmetto, the interior is so light compared with the hard epidermis that the result is practically a hollow cylinder.

1. Old Fort Moultrie near Charleston was built originally of palmetto logs; was this good engineering or not? Why?

2. Why is a stalk of sugar cane so much heavier than one of corn? A green cornstalk than a dead one? (215.)

3. Explain the advantages of structure in a culm of wheat; a stalk of corn; a reed. (218.)

4. Would the same quality be of advantage to an oak? Why, or why not?

5. Is it any advantage to the farmer that grain straw is so light?

STEMS OF DICOTYLEDONS

MATERIAL. — Twigs from one to three years old of almost any kind of hard wood shoots; elm, basswood, mulberry, leatherwood, and pawpaw show the bast well; sassafras, slippery elm, birthwort (*Aristolochia*), and in spring, hickory and willow, show the cambium; grape and Trumpet vine the ducts. Have some twigs placed in red ink from four to twelve hours before the lesson begins. Grape, peach, or hickory will answer well for this purpose.

219. Examination of a Typical Specimen. — Examine carefully the outer surface of a young twig, not less than

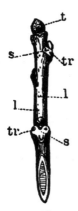

one nor more than three years old, of any convenient specimen. Notice the scars left by the leaves of the season as they fell away, and look for one or more little roundish dots called *leaf traces*, that mark the points where the fibrovascular bundles from the leaf veins passed into the stem. The little oblong or lens shaped corky spots that dot the surface of a twig are called *lenticels*. They are the breathing pores or ventilators through which the air penetrates to the inner parts of the stem. They usually disappear on older branches, where the outer bark is constantly breaking away and sloughing off. Sometimes, however, they are quite persistent, as in the peach, cherry,

299. — Alternate leaved twig of walnut: *t*, terminal bud; *s,s*, leaf scars; *tr*, leaf traces; *l,l*, lenticels.

and china tree. The characteristic markings of the birch bark, which make it so ornamental, are due to the lenticels. As the tree grows they elongate either vertically, by the lengthening of the twig, or horizontally, by its increase in diameter, until they often appear as long slits.

Scrape off a little of the brownish, or sometimes almost colorless outer covering. This is the epidermis, and is replaced by the outer corky layer of the bark in older stems. As the stem increases in diameter from year to year this outer covering is broken up and pushed aside to make way for the new growth, so that the bark is constantly dying and sloughing off from the outside and as constantly renewed from within. Under the epidermis, notice a greenish layer of young bark; beneath this a layer of rather tough, stringy fibers called *bast*, and finally a harder woody substance that constitutes the bulk of the interior of the stem. Cut through this to the very center of the axis and we find a cylinder of lighter, pithy texture; this is the same as the parenchyma or parent tissue that we found pervading the interior of the cornstalk (Sec. 213). It is usually called the pith or *medulla*, and is the only part present in very young stems.

Between the woody axis and the bark is a more or less soft and juicy ring called

220. The Cambium Layer. — This is not always easily distinguishable with a hand lens, but is conspicuous in the stems of sassafras, slippery elm, aristolochia, etc. If some of these can not be obtained, the presence of the cambium can be recognized by observing the tendency of most stems to "bleed" when cut, between the wood and bark. This is because the cambium is the active part of the stem in which growth is taking place, and consequently it is most abundantly supplied with sap. This is especially the case in spring, when it becomes so gorged with nourishment that if a rod of hickory or elder is pounded, the pulpy cambium is broken up and the bark may be slipped off whole from the wood. It is the nourishment contained in the cambium of certain plants that tempts goats and calves to bark them in spring, and that enables savages, in time of dearth, to subsist for a while on the buds and bark of trees.

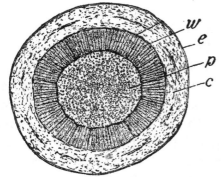

221. **Difference between Dicotyledons and Monocotyledons.** — Cut cross and vertical sections of your specimen, and sketch them as seen under the lens, labeling the different parts that have been examined. Refer to Figures 300 and 301 if you have any difficulty in distinguishing the parts. Notice the little pores or cavities that dot the woody part in the cross section; where are they largest and most abundant? How are the rings marked off from one another? These pores are sections of the ducts already alluded to (Secs. 214, 215). They are very large in the grape vine, and a cutting two or three years old will show them distinctly. Examine cross and vertical sections of a twig that has stood in red ink from three to twelve hours and observe the course the fluid has taken. (The rapidity with which the liquid is absorbed varies with different stems and at different seasons. It is

300. — Section across a young twig of box elder, showing the four stem regions: *e*, epidermis, represented by the heavy bounding line; *c*, cortex; *w*, vascular cylinder; *p*, pith. (From COULTER'S "Plant Relations.")

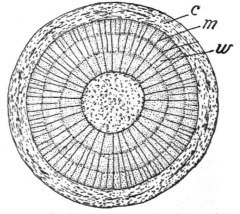

301. — Section across a twig of box elder three years old, showing three annual growth rings, in the vascular cylinder. The radiating lines (*m*), which cross the vascular region (*w*), represent the pith rays, the principal ones extending from the pith to the cortex (*c*). (From COULTER'S "Plant Relations.")

most rapid in spring and slower in winter. In grape, plum, and peach it ascends quickly.) What should you infer from this as to the office of the ducts? How does this conclusion compare with your observations on the vascular bundles of monocotyledonous stems? Notice that the dicotyledon differs from monocotyledonous stems in having the pith all gathered in a narrow cylinder in the center, and the vascular tissue arranged in one or more concentric layers around it, according to the age of the stem. In general, dicotyl stems may be said to include four regions; 1st, the epidermis or bark, *e* (Fig. 300); 2d, the cortex, *c*, made up of the cambium and bast, with certain other tissues; 3d, the vascular cylinder, or woody portion *w*, made up of concentric rings each representing a year's growth; and 4th, the pith *p*, medulla, or parenchyma, as it is variously termed by botanists.

222. Medullary Rays.—Observe the whitish silvery lines that radiate in every direction from the center, like the spokes of a wheel from the hub. These are the medullary rays and consist of threads of pith that serve as lines of communication between the "parent tissue" and the growing cambium layer. In old stems the central pith frequently disappears and its office is filled by the medullary rays, which become quite conspicuous.

223. The Rings, into which the vascular cylinder is divided, mark the yearly additions to the growth of the stem, which increases by the constant addition of fibro-vascular bundles from the outside; hence such stems are called *exogens* or "outside growers."

224. The Structure of the Fibro-vascular Bundles is somewhat complicated and can not be studied to advantage without the aid of a compound microscope, but a little attention to the diagrams will make it intelligible. The inner part of each bundle (*i.e.*, the part toward the axis) is made up of woody fibers shown at

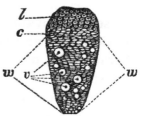

302.—Transverse section of vascular bundle from stem of a dicotyledon: *l*, bast; *c*, cambium; *v*, ducts, *w*, wood cells.

w (Fig. 302), intermingled with larger sized tubes or ducts, *v*, the sections of which made the pores referred to in Section 221. In front of these is the cambium layer, *c*, and

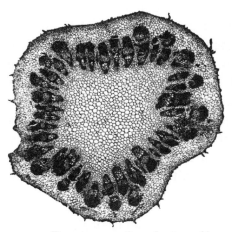

beyond that, the soft bast and other tissues in which elaborated food is being brought down from the leaves and material for growth provided. In very young stems the vascular bundles are separate and distinct, as in Figure 303, being connected only by a ring of cambium, but as growth advances and more bundles are formed to supply the new buds and leaves of the devel-

303. — Transverse section of a stem of burdock, showing fibrovascular bundles not completely united into a ring.

oping axis, they become crowded into a ring (Fig. 304), which is separated into woody wedges by the threads of pith (medullary rays) that run between them from the center to the cortex. The cambium constantly advances outwards, beginning every spring a new season's growth and leaving behind the ring of ducts and woody fibers made the year before. As the work of the plant is most active and its growth most vigorous in spring, the largest ducts are formed then, the tissue becoming closer and finer as the season advances, thus causing the division into annual rings that is so characteristic of dicotyl stems. Each new stratum of growth is made up of the fibrovascular bundles that supply

304. — Diagram of an older dicotyl stem, showing bundles confluent into a ring (GRAY).

the leaves and buds and branches of the season. Figure 305 gives a diagrammatic section illustrating the passage of the bundles from the leaves to the stem of a dicotyledon, each successive node sending down its quota.

In this way we see that the increase of dicotyl trunks and branches is approximately in an elongated cone (Fig. 306), the number of rings gradually diminishing toward the top till at the terminal bud of each bough it is reduced to

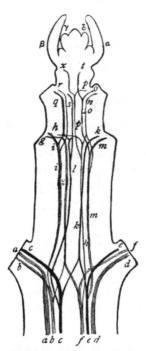

305. — Diagrammatic view of a leafy stem of clematis, showing the arrangement of the fibrovascular bundles: *a, b, c, —e, f, d,* the fascicles from the lower pair of leaves; *i, g, l, —k, h, m,* the fascicles from the second pair of leaves; *q, r, s, —p, n, o,* the fascicles from the third pair of leaves; *x, t,* fascicles of the fourth pair of leaves; β. *a, —γ, δ,* pairs of undeveloped leaves not as yet having fascicles (GRAY, *after* NÄGELI).

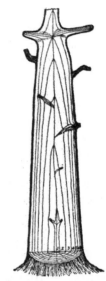

306. — Diagram illustrating the annual growth of dicotyledons.

a single one, as in the stems of annuals.

Sometimes a late autumn, succeeding a very dry summer, will cause trees to take on a second growth, and thus form two layers of wood in a single season, so we can not always rely absolutely upon the number of rings in estimating the age of a tree.

225. The Stems of Conifers. — Examine a young stem of pine, and compare with the one just studied. What difference do you notice? This absence of the duct pores constitutes one of the most conspicuous differences between the stems of conifers (cone bearers) and dicotyledons.

The ducts are there, but they are formed differently from those of other exogens, and can not be studied without a compound microscope. From what part of the stem does the rosin exude? Place a cutting in red ink and notice through what part the fluid rises; where, would you judge from this, is the most active part of the stem?

PRACTICAL QUESTIONS

1. Explain the principle upon which boys slip the bark from certain kinds of wood in spring to make whistles. (220.)

2. Why can not they do this in autumn or winter?

3. Name some of the plants commonly used for this purpose.

4. Is the spring, after the buds begin to swell, a good time to prune fruit trees and hedges? Why? (220.)

5. What is the best time, and why?

6. Why are grape vines liable to bleed to death if pruned too late in spring? (220, 221.)

7. Why are nurserymen, in grafting, so careful to make the cambium layer of the graft hit that of the stock? (220.)

8. In calculating the age of a tree or bough from the rings of annual growth should we take a section from near the tip, or the base? Why? (224.)

MOVEMENT OF WATER THROUGH THE STEM

MATERIAL. — An egg, a small cup, and some salt water. A potted young plant of corn, calla lily, tropæolum, sunflower, etc. A few centimeters each of glass tubing and rubber tubing about the diameter of the stem of the plant. A twig of willow, currant, or other easily rooting shrub.

226. Difficulty of Accounting for Sap Movement. — Just what causes the rise of sap in the stem is one of the puzzles of vegetable physiology that botanists have not yet been able to solve completely. It is closely connected with the phenomena of transpiration, the rapidity of the current increasing and decreasing according to the activity of the evaporating surfaces. If loss of water begins at any spot through growth or transpiration, the nearest tissues give up their water first, then the more remote, and so on, till the most distant — generally the roots — have to absorb water from without, and thus a constant current is kept up toward the places where moisture is needed.

227. Osmose. — The rise of sap is partly due to the pressure caused by the constant absorption of soil water through the absorbent hairs of the root. The passage of liquids through the walls of cells and tissues is known as *osmose* and takes place when liquids of different densities are separated by a thin membrane, the principle governing the direction of the flow being that the thinner, lighter liquid passes toward the denser. The nature of the substances, also, must be considered; those that are crystalline and easily soluble, like sugar and salt, pass readily through membranes, while gelatinous ones pass with difficulty or not at all.

Chip away a bit of the shell from the big end of an egg, taking care not to injure the thin membrane underneath. Make a small puncture through both shell and membrane in the small end and place the egg in a cup with its big end in salt water. In a few hours the contents will be found running out of the puncture at the other end, having been forced out by the water that made its way in below. And there are no pores visible, even with the most powerful microscope, in the membrane that lines the eggshell.

The same principle is well illustrated by the experiment described in Section 204, the water passing by osmose through the walls of the cells that make up the substance of the stem. Take one of the stem sections after it has lain in fresh water, and transfer it to a five per cent solution of salt water (about a tablespoonful of salt to a tumbler of liquid). Allow it to remain as before, and then examine. It will be found to have become straight again, or perhaps even to have coiled over in the opposite direction. This is because the thinner liquid of the cells has passed out by osmose into the thicker salt solution, so that the interior cells have become flabby, while the exterior ones, protected by the epidermis, remain distended and thus cause the section to curve inward.

The passage of liquids into a sac or cell is called *endosmose*, out of it, *exosmose*. Which is it that takes place between the soil water and the root?

228. Action of Osmose in the Root. — The sap within the root is generally denser than the water of the soil, so there is a continuous osmotic flow from the latter to the former, but within the stem the fluid is more nearly of the same density throughout and the conditions for osmosis are not so favorable, though it probably does take place to some extent. A more efficient cause is generally held to be the force exerted by the upward pressure of water absorbed into the roots, and known as

229. Root Pressure. — Cover a calla lily, young cornstalk, sunflower, or other succulent herb with a cap of oiled paper to prevent transpiration, set the pot containing it in a pan of warm water and keep it at a gentle heat. After a few hours look for water drops on the leaves. Where did this water come from? How did it get up into the leaves?

Now cut off the stem of the plant six or eight centimeters (three or four inches) from the base. Slip over the part remaining in the soil a bit of rubber tubing of about the same diameter as the stem, and tie tightly just below the cut. Pour in a little water to keep the stem moist, and slip in above a short piece of tightly fitting glass tubing. Watch the tube for several days and note the rise of water in it. The same phenomenon may be observed in the "bleeding" of rapidly growing, absorbent young shoots, such as grape, sunflower, gourd, tobacco, etc., if cut off near the ground in spring when the earth is warm and moist. This flow can not be due to transpiration, since the leaves and other transpiring parts have been removed. Transpiration, by causing a deficiency of moisture in certain places may influence the direction and rapidity of the current, but does not furnish the motive power, which evidently comes, in part at least, from the roots, and is the expression of their absorbent activity.

230. Root Pressure and Root Pull. — There is no antagonism between these two forces. Root pull affects the body of the plant with its system of tubes and cells ; root pressure affects the free contents of these parts, just as we may sink a water pipe into the ground and at the same time force the water upward through it.

231. Direction of the Current. — Remove a ring of the cortical layer from a twig of any readily rooting dicotyledon, being careful to leave the woody part with the cambium intact. Place the end *below* the cut ring in water, as shown in Figure 307. The leaves above the girdle will remain fresh. How is the water carried to them? How does this agree with the movement of red ink observed in Section 221?

Next prune away the• leaves and protect the girdled surface with tin foil, or insert it below the neck of a deep bottle to prevent evaporation and wait until roots develop. Do they come most abundantly from above or below the decorticated ring?

These experiments show that the upward movement of crude sap toward the leaves is mainly through the ducts in the woody portion of the stem, while the

307. — A twig which had been kept standing in water after the removal of a ring of cortical tissue: a, level of the water; b, welling formed at the upper denudation; c, roots.

downward flow of elaborated sap from the leaves takes place chiefly through the soft bast and certain other vessels of the cortical layer.

232. Ringing Fruit Trees. — This explains why farmers sometimes hasten the ripening of fruit by the practice of ringing. As the food material cannot pass below the denuded ring, the parts above become gorged and a process of forcing takes place. The practice, however, is not to be commended, except in rare cases, as it generally leads to the death of the ringed stem. The portion below the ring can receive no nourishment from above, and will gradually be so starved that it can not even act as a carrier of crude sap to the leaves, and so the whole bough will perish.   Figure 308 will give a good general idea of the movement of sap in trees, the arrows indicating the direction of the movement of the different substances.

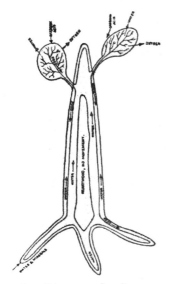

308. — Diagram showing general movement of sap.

233. Sap Movement not Circulation. — It must not be supposed that this flow of sap in plants is analogous to the circulation of the blood in animals, though frequently spoken of in popular language as the "circulation of the sap." There is no central organ like the heart to regulate its flow, and the water taken up by the roots does not make a continual circuit of the plant body as the blood does of ours, but is dispersed by a process of general diffusion, part into the air through transpiration, and part through the plant body as food, wherever it is needed.

234. Unexplained Phenomena. — While root pressure will account for the rise of sap to a certain extent, none of the causes assigned by physiologists are sufficient to explain all the phenomena. The highest force as yet proved to be exerted by it is sufficient to balance a column of water only ten to fifteen meters (thirty to fifty feet) high. The power with which it acts seems to vary in different plants. In the nettle it is capable of lifting the sap to a height of about 4.5 meters (15 feet) and in the grapevine more than 11 meters, or about 36.5 feet. It is claimed that in the birch it exerts a lifting force nearly equal to the pressure of a column of water eighty-five feet high, but even this is quite inadequate to explain the rise of sap to the tops of trees three hundred and four hundred feet high, like the giant redwoods of California or the still taller blue gums of Australia. Capillary attraction and the buoyant force of air bubbles in the cavities of the stem, in conjunction with various other causes, have been called in to explain the phenomenon, but so far as our knowledge goes at present none of them seems to account for it satisfactorily.

PRACTICAL QUESTIONS

1. In pruning, why should the cutting be confined as far as possible to young shoots?

2. Why should vertical shoots be cut off obliquely?

3. Why should pruning not be done in wet weather?

4. Why will a leafy shoot heal more quickly than a bare one? (24, 25, 26, 200.)

5. Why does a transverse cut heal more slowly than a vertical one? (231, 232.)

6. Why does a ragged cut heal less readily than a smooth one?

7. Why does the formation of wood proceed more rapidly as the amount of transpiration is increased? (226.)

8. Why do nurserymen sometimes split the cortex of young trees in summer to promote the formation of wood? (219.)

9. What is the advantage of scraping the stems of trees?

10. Explain the frothy exudations that often appear at the cut ends of firewood, and the singing noise that accompanies it. (215, 224.)

11. What advantage is it to high climbing plants, like grape and trumpet vine (*Tecoma*), to have such large ducts? (214, 215, 221.)

12. Why is the process of layering more apt to be successful if the shoot is bent or twisted at the point where it is desired to make it root?

13. Why do oranges become dry and spongy if allowed to hang on the tree too long? (215, 231, 232.)

14. Why will corn and fodder be so much richer in nourishment if, instead of pulling the fodder when it is mature and leaving the ears to ripen in the field, we cut down the whole stalk and allow both fodder and grain to mature upon it? (215, 231, 233.)

15. Why will inserting the end of a wilted twig in warm water sometimes cause it to revive? (229.)

16. Why should we protect the south side rather than the north side of tree trunks in winter?

17. Why does cotton run all to weed in very wet weather?

18. Why in pruning a branch is it best to make the cut just above a bud?

19. Why is the rim of new bark or callus that forms on the upper side of a horizontal wound thicker than that on the lower side? (231.)

20. Why is it that the medicinal or other special properties of plants are found mostly in the leaves and bark, or parts immediately under the bark? (215, 220.)

WOOD STRUCTURE

MATERIAL. — Select from the billets of wood cut for the fire, sticks of various kinds; hickory, ash, oak, chestnut, maple, walnut, cherry, pine, cedar, tulip tree, all make good specimens. Red oak shows the medullary rays particularly well. Get sticks of green wood if possible and have them planed smooth at the ends. It would be well for the teacher to have a hatchet and let the class collect their own specimens. Collect also, where they can be obtained, waste bits of dressed lumber from a carpenter or joiner. For city schools prepared samples should be obtained of the dealers. If nothing better is available, any pieces of unpainted woodwork about the schoolroom will furnish subjects for study.

309. — Cross section through a black oak, showing heartwood and sapwood
(from PINCHOT, U. S. Dept. of Agr.).

235. Detailed Structure of a Woody Stem. — Select a good-sized billet of hard wood and count the rings of annual growth. How old was the tree or the bough from which it was taken? Was its growth uniform from year to year? How do you know? Are the rings broadest, as a general thing, toward the center or the circumference? How do you account for this? Is each separate ring of uniform thickness all the way round? Mention some of the circumstances that might cause a tree to grow less on one side than on the other; such, for instance, as too great shading, lack of foliage development from one cause or another, exposure of roots by denudation, etc. Are the rings of the same thickness in all kinds of wood? Which are the most rapid growers, those with broad or with narrow rings? Do you notice any difference in the texture of the wood in rapid and in slow growing trees? Which makes the better timber as a general thing, and why?

310. — Vertical section through a black oak (from PINCHOT, U. S. Dept. of Agr.).

236. Heartwood and Sapwood. — Notice that in some of your older specimens (cedar, black walnut, barberry, black locust, chestnut, oak, Osage orange, show the difference distinctly) the central part is different in color and texture from the rest. This is because the sap gradually abandons the center (Sec. 224) to feed the outer layers where growth in dicotyls takes place; hence, the outer part of the stem usually consists of sapwood, which is soft and worthless as timber, while the dead interior forms the

durable heartwood so prized by lumbermen. The heartwood is useful to the plant principally in giving strength
and firmness to the axis. It will now be seen why gird-
ling a stem, that is, chipping off a ring of the softer parts

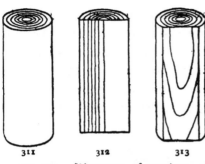

all round, will kill it, while
we often see vigorous and
healthy trees with the center of the trunk entirely
hollow.

311   312   313

311–313. — Diagrams of sections of
timber: 311, cross section; 312, radial;
313, tangential (from PINCHOT, U. S.
Dept. of Agr.).

**237. Vertical Arrange-
ment.** — In studying the
vertical arrangement of
stems two sections are necessary, a radial and a tangential one. The former
passes along the axis, split-

ting the stem into halves (Fig. 312); the latter cuts
between the axis and the perimeter, splitting off a segment
from one side (Fig. 313).

238. The Graining of Timber. — It is the medullary rays
that constitute the characteristic graining of different
woods. In a chip of red oak or chestnut from just beneath

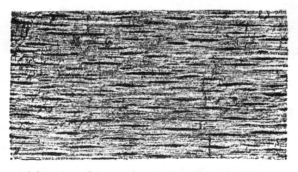

314. — Tangential section of mountain ash, showing ends of the medullary rays.

the bark their cut ends can be seen very distinctly with the
naked eye. Split a thicker chip of the same kind parallel
with the medullary rays and notice the difference, the
rays now appearing as silvery bands traversing the wood.

Compare the graining of your specimens, or of the floor-
ing, window casings, doors, desks, benches, etc., of your
schoolroom with Figures 312 and 313, and tell what kind

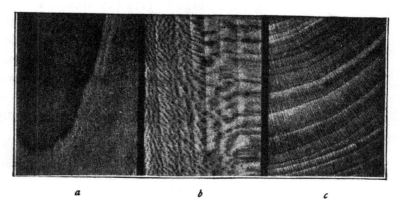

a   b   c

315. — Sections of sycamore wood (from PINCHOT, U.S. Dept. of Agr.) :
a, tangential; b, radial; c, cross.

of cut was made in each case and show how the appearance
of the timber has been affected by it.

239. Knots. — Look for a billet with a knot in it. Notice
how the rings of growth are disturbed and displaced in its
neighborhood. If the knot is a large one, it will itself

316. — Sections of white pine wood (from PINCHOT, U.S. Dept. of Agr.).

have rings of growth. Count them, and tell what its age
was when it ceased to grow. Notice where it originates.
Count the rings from its point of origin to the center of
the stem. How old was the tree when the knot began to
form? Count the rings from the origin of the knot to the

circumference of the stem; how many years has the tree lived since the knot was formed? Does this agree with

317. — Section of branch showing knot.

the age of the knot as deduced from its own rings? (As the tree may continue to live and grow indefinitely after the bough which formed the knot died or was cut away, there will probably be no correspondence between the two sets of rings, especially in the case of old knots that have been covered up and embedded in the wood.)

The longer a dead branch remains on a tree the more rings of growth will form around it before covering it up, and the greater will be the disturbance caused by it. Hence, timber trees should be pruned while very young, and the parts removed should be cut as close as possible to the main branch or trunk. Sometimes knots injure lumber very much by falling out and leaving the holes that are so often seen in pine boards. In other cases, however, when the knots are very small, the irregular markings caused by them

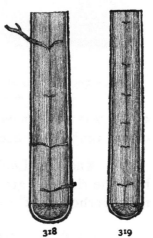

318   319

318, 319. — Diagrams of tree trunks, showing knots of different ages: 318, from tree grown in the open; 319, from tree grown in a dense forest.

add greatly to the beauty of the wood. The peculiar marking of bird's-eye maple is caused by abortive buds buried in the wood.

PRACTICAL QUESTIONS

1. Name the principal timber trees of your neighborhood. What gives to each its special value?

2. Which is better for timber, a tree grown in the open, or one in a forest, and why? (239.)

3. What are the objects to be attained in pruning timber trees? Orchard and ornamental trees?

320. — Timber tree spoiled by standing too much alone in early youth (from PINCHOT, U. S. Dept. of Agr.). Notice how the crowded young timber in the background is righting itself, the lower branches dying off early from overshading, leaving tall, straight, clean boles.

4. What is the difference between timber and lumber? Between a plank and a board? Between a log, stick, block, and billet?

5. Is the outer bark of any use to a tree, and if so, what? (176, 191, 219.)

6. Why does sapwood decay more quickly than heartwood?

FIELD WORK

Make a study of the various climbing plants of your neighborhood with reference to their modes of ascent, and the effect, injurious, or other, upon the plants they cling to. Note the direction of twining stems and tendrils, and their various adaptations to their office. Consider whether the twining habit might not lead to parasitism, especially in the case of soft-stemmed twiners when brought into contact with soft-stemmed annuals. Observe the various habits of stem growth; prostrate, declined, ascending, etc., and see what adaptation to circumstances can be detected in each case.

Notice the shape of the different stems met with, and learn to recognize the forms peculiar to certain of the great families. Observe the various appliances for defense and protection with which they are provided, and try to find out the meaning of the numerous grooves, ridges, hairs, prickles, and secretions that are found on stems. Always be on the alert for transformations, and learn to recognize a stem under any disguise, whether thorn, tendril, foliage, water holder, etc.

Note the color and texture of the bark of the different trees you see, and learn to distinguish the most important by it. Observe the difference in texture and appearance of the bark on old and young boughs of the same species. Try to account for the varying thickness of the bark on different trees and on different parts of the same tree. Farmers are generally engaged in clearing and pruning at this season, and it will probably not be difficult to get all the specimens needed among the rubbish they are clearing away. Notice the difference in the timber of the same species when grown in different soils, at different ages of the tree, and in healthy and weakly specimens.

VII. BUDS AND BRANCHES

BRANCHING STEMS

MATERIAL. — Twigs of hickory and buckeye, or other alternate and opposite leaved plants with well-developed terminal buds. A larger bough of each should also be provided, and where practicable, twigs of several different kinds for comparison. Lilac, horse-chestnut, maple, ash, viburnum, are good examples of opposite buds.

240. Modes of Branching. — Compare the arrangement of the boughs on a pine, cedar, magnolia, etc., with those of the elm, maple, apple, or any of our common deciduous trees. Draw a diagram of each showing the two modes of growth. The first represents the *excurrent* kind, from the Latin *excurrere*, to run out; the second, in which the trunk seems to divide at a certain point and flow away and lose itself in the branches, is called *deliquescent*, from the Latin *deliquescere*, to melt or flow away. The great majority of stems, as a little observation will show, present a mixture of the two modes.

321. — Diagram of ex-current growth.

322. — Diagram of deliquescent growth.

241. Terminal and Axillary Buds. — Notice the large bud at the end of a twig of hickory, sweet gum, beech, cottonwood, etc. This is called the *terminal* bud because it terminates its branch.. Notice the leaf scars on your twig, and look for the small buds just above them.

These are *lateral*, or *axillary* buds, so called because. they spring from the axils of the leaves. How many leaves did your twig bear? How many ranked? What difference in size do you notice between the terminal and lateral buds?

323. — Young bud of hickory (*after* GRAY): *t*, terminal bud; *rs*, ring of scars left by bud scales of previous years; *s*, leaf scars; *l,l*, lenticels; *tr*, leaf traces.

242. The Leaf Scars. — Examine the leaf scars with a hand lens, and observe the number and position of the little dots in them. (Ailanthus, varnish tree, and china tree show these very distinctly.) Refer to Section 219, and say what these dots are.

243. Bud Scales and Scars. — Notice the stout, hard scales by which all the buds are covered. Pull these away from the terminal one and notice the ring of scars that they leave around the base of the bud. Look lower down on your twig for a ring of similar scars left from last year's bud. Is there any difference in the appearance of the bark above and below this ring? If so, what is it, and how do you account for it? Is there more than one of these rings of scars on your twig, and if so, how many? How old is the twig and how much did it grow each year? Has its growth been uniform or did it grow more in some years than others?

244. Different Rates of Growth. — Notice the very great difference between branches in this respect. Sometimes the main axis of a shoot will have lengthened from twenty to fifty centimeters (eight to twenty inches) or more in a single season, while some of the lateral ones will have grown but an inch or two in four or five seasons. One reason of this is because the terminal bud, being on one of the great trunk lines of sap movement, gets a larger share of nourishment than the rest, and being stronger and better developed, starts out in life with superior advantages of position. Then, too, in ordinary upright stems the sap flow is strongest in the upper part of the stem, as may be shown by selecting two healthy seedlings as nearly as may be of the same size and height, inverting one of them as described in Section 159, and keeping it in this position for several days by tying or by attaching a weight to it, while leaving the other upright. Watch their growth for a week or ten days and note results.

Make a drawing of your specimen, showing all the points brought out in the examination just made. Cut sections above and below a set of bud scars and count the rings of annual growth in each section. What is the age of each? How does this agree with your calculation from the number of scar rings?

245. Irregularities. — Take a larger bough of the same kind that you have been studying, and observe whether the arrangement of branches on it corresponds with the arrangement of buds on the twig. Did all the buds develop into branches? Do those that did develop all correspond in size and vigor? If all the buds developed, how many branches would a tree produce every year?

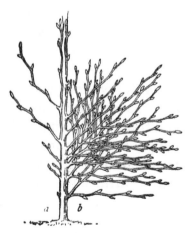

324.—Bud development of beech: *a*, as it is, many buds failing to develop; *b*, as it would be if all the buds were to live.

In the elm, linden, beech, hornbeam, hazelnut, willow, and various other plants, the terminal bud always dies and the one next in order takes its place, giving rise to the more or less zigzag axis that generally characterizes trees of these species.

246. Forked Stems. — Take a twig of buckeye, horse-chestnut, or lilac, and make a careful sketch of it, show-

ing all the points that were brought out in the examination
of your previous specimen. Which is the larger, the lateral

or the terminal bud? (If lilac is used, there
will probably be no terminal bud.) Is their
arrangement alternate or opposite? What
was the leaf arrangement? Count the dots
in the leaf scars; are they the same in all?
If all the buds had developed into branches,
how many would spring from a node?
Look for the rings of scars left by the last
season's bud scales. Do you find any twig
of more than one year's growth, as measured
by the scar rings?

325.—Opposite-
leaved twig of
horse-chestnut.

Look down between the forks of a
branched stem for a round scar. This is not a leaf scar,
as we can see by its shape, but one left by the last season's
flower cluster. The flower, as we all know, dies after
perfecting its fruit, and so a flower bud can not continue
the growth of its axis, as other buds do, but has just the
opposite effect and stops all further growth in that direction. Hence, stems and branches that end in a flower
bud can never develop either excurrent
or ordinary deliquescent growth, but
are characterized by short branches
and frequent forking. The same thing
happens when, for any reason, the
terminal bud is destroyed or injured
either artificially, or through natural
processes, as in the lilac, where it
is frequently aborted and its place
usurped by the two nearest lateral
ones, which put forth on each side of
it and continue the growth of the

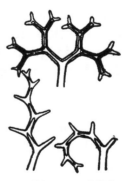

326.— Diagrams of dichot-
omous branching.

branch in two forks instead of a single axis. This gives
rise to the kind of branching which we see exemplified in
the lilac, buckeye, horse-chestnut, dogwood, jimson weed,
etc., designated by botanists as *dichotomous*, or two-forked.

Draw a diagram of the buckeye, or other dichotomous

stem as it would be if all the buds developed into branches, and compare it with your diagrams of excurrent and deliquescent growth.

247. Definite and Indefinite Annual Growth. — The presence or absence of terminal buds gives rise to another important distinction in plant development — that of *definite* and *indefinite* annual growth. Compare with any of the twigs just examined, a branch of rose, honey locust, sumac, mulberry, etc., and note the difference in their modes of termination. The first kind, where the bough completes its season's increase in a definite time and then devotes its energies to developing a strong terminal bud to begin the next year's work with, are said to make a *definite or determinate annual growth*. Those plants, on the other hand, which make no provision for the future but go straight on flourishing and rejoicing, like the grasshopper in the fable, till the cold comes and literally nips them in the bud, are *indefinite*, or *indeterminate* annual growers. Notice the effect of this habit upon their mode of branching. The buds toward the end of each shoot, being the youngest and tenderest, are most readily killed off by frost or other accident, and hence the new branches spring mostly from the older and stronger buds near the base of the stem. It is this mode of branching that gives to plants of this class their peculiar bushy aspect. Such shrubs generally make good hedges on account of their thick undergrowth. The same effect can be produced artificially by pruning.

248. Differences in the Branching of Trees. — We are now prepared to understand something about the causes of that endless variety in the spread of bough and sweep of woody spray that makes the winter woods so beautiful.

327. — Winter spray of ash, an opposite-leaved tree.

Where the terminal bud is undisputed monarch of the bough, as in the pine and fir, or where it is so strong and vigorous as to overpower its weaker brethren and

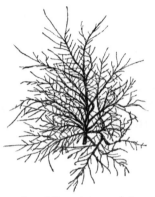

keep the lead, as in the magnolia and holly, we have excurrent growth. In plants like the oak and apple, on the other hand, where all the buds have a more nearly equal chance, the lateral branches show more vigor and the result is either deliquescent growth, or a mixture of the two kinds. In the elm and beech, where the usurping pseudo-terminal bud keeps the mastery, but does not completely

328. — Winter spray of elm.

overpower its weaker brethren, we find the long, sweeping, delicate spray characteristic of those species. Examine a sprig of elm and notice further that the flower buds are all down near the base of the stem, while the leaf buds are near the tip. The chief development of the season's growth is thus thrown toward the end of the branch, giving rise to that fine, feathery spray which makes the elm an even more beautiful object in winter than in summer.

An examination of the twigs of other trees will bring out the various peculiarities that affect their mode of branching. The angle, for instance, which a twig makes with its bough has a great effect in shaping the contour of the tree. As a general thing, acute angles produce slender, flowing effects ; right, or obtuse angles, more bold and rugged outlines.

PRACTICAL QUESTIONS

1. Has the arrangement of leaves on a twig anything to do with the way a tree is branched? (68, 241.)

2. Why do most large trees tend to assume the excurrent, or axial mode of growth if let alone? (244.)

3. If you wished to alter the mode of growth, or to produce what nurserymen call a low-headed tree, how would you prune it? (246, 247.)

4. Would you top a timber tree? (246, 247.)

5. Are low-headed or tall trees best for an orchard?

6. Why is the growth of annuals generally indefinite?

7. Name some trees of your neighborhood that are conspicuous for their graceful winter spray.

8. Name some that are characterized by the sharpness and boldness of their outlines.

9. Account for the peculiarities in each.

BUDS

MATERIAL. — Expanding buds of any of the kinds used in Sections 240–248 and of tulip tree, magnolia, or other plant with stipular leaf scales. The buds should be in different stages of development, some of them partly expanded. Beech, elm, oak, sycamore, hackberry, fig, will any of them serve as examples of stipular scales, but it is advisable always to use the largest buds obtainable. City schools might get a young India rubber tree from a nursery, or buds of cultivated magnolia from a florist. Gummy buds like horse-chestnut and Lombardy poplar should be soaked in warm water before dissecting, to soften the gum. Buds with heavy fur on the scales, or on the parts within them, can not very well be studied in section, but the parts must be taken out and examined separately. Where material is scarce, the twigs used in Sections 240–248 can be placed in water and kept until the buds begin to expand.

249. Study of an Opposite-Leaved Bud. — Examine a twig of buckeye, horse-chestnut, lilac, or maple, etc., just as the buds are beginning to unfold. Make an enlarged sketch of the terminal one (in the lilac, usually two), showing the relative size and position of the scales.

250. Arrangement of the Scales. — Notice the manner in which the scales overlap, so as to break joints, like shingles on the roof of a house. Leaves or scales that overlap in this way are said to be *imbricated*. Where the

329. — Diagram of opposite bud scales.

leaves are opposite, as in the specimen we are examining, the manner of imbrication is very simple. Remove the scales one by one, representing the number and position of the pairs by a diagram after the model given in Figure 329. (If the scales are too brittle to be removed without breaking, use a bud that has been soaked in warm water for an hour or two.) How many pairs of scales are there in each set? How does their arrangement correspond with that of the leaf scars upon the stem? What difference in size and texture do you observe between the outer and inner scales?

330. — Development of the parts of the bud in the buckeye (*after* GRAY).

251. Nature of the Scales. — Hold up to the light one of the scales from a partly expanded bud and see whether it is veined, and in what way. Does this correspond with the venation of foliage leaves? Can you make out what the scales represent? Their arrangement is the same as that of the leaves, so they must represent the leaf or some part of it, as the petiole or the stipules. In the lilac and various other buds they are found in all stages of transition from scales to true leaves, from which their real nature may readily be inferred. In the common buckeye and the horse-chestnut the transition is not so apparent, but a comparison with Figure 330 will show that they are altered petioles.

252. Use of the Scales. — What purpose do the scales serve? You can best answer this question by asking yourself what is the use of the shingles on the roof of a house, or of the cloaks with which we wrap ourselves in winter? Notice how thick and hard the outer ones are, and how the inner ones envelop the tender parts within like blankets. As we sometimes coat our roofs with tar and cement, so these scales, especially in cold climates, are often coated with gum for greater security against the weather.

253. Internal Structure of the Bud. — Make a cross section of a bud and sketch it as it appears under the lens.

Next draw a vertical section, then remove the contents and see what they are. There will be no difficulty in recognizing the circle of young leaves just within the scales. How many of these rudimentary leaves are there? Is their arrangement alternate or opposite? Notice the down with which they are covered (in the horse-chestnut and buckeye). Have the mature leaves of these plants any covering of this sort? What is its use here?

254. Folding of the Leaves. — Notice the manner in which the young leaves are folded in the bud. This is called by botanists *vernation*, or *prefoliation*, words meaning respectively " spring condition " and " condition preceding the leaf." Leaves have to be packed in the bud so as to occupy the least space possible, and in different plants they will be found folded in a great many different ways, as is best suited to the shape and texture of the leaf and the space available for it

331, 332. — Buds of maple: 331, vertical section of a twig; 332, cross section through an end bud, showing folded leaves in center and scales surrounding them.

in the bud. When doubled back and forth like a fan, or crumpled and folded as in the buckeye, horse-chestnut, and maple, the vernation is *plicate* (Fig. 332).

255. Position of the Flower Cluster. — What do you find within the circle of leaves? Examine one of the smaller axillary buds, and see if you find the same object within it. If you are in any doubt as to what this object is, examine a bud that is more expanded and you will have no difficulty in recognizing it as a rudimentary flower cluster. Notice its position with reference to the scales and leaves. Being at the center of the bud, it will, of course, terminate its axis when the bud expands, and the growth of the branch will culminate in the flower. The branching of the buckeye (or horse-chestnut) must, then, be of what order?

Compare your drawings with the section of a hyacinth bulb or jonquil, and note the similarity in position of the flower clusters.

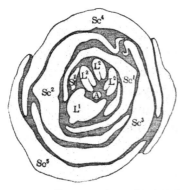

333. — Cross section of a leaf bud of the rose, showing the alternate arrangement of scales and rudimentary leaves: A, growing point; L^1, youngest leaf; L^2, three folded lobes of second leaf; St^2, stipules of second leaf; Sc^1-Sc^5, scales.

256. Study of an Alternate-Leaved Bud. — Examine a large terminal bud of hickory, just about to open. (Apple, pear, cherry, etc., may be substituted if necessary.) How do the scales differ in shape and texture from those already examined? Pick off the scales one by one, noting their position carefully and illustrating it by a diagram, as shown in Figure 333. This is another variety of the imbricated arrangement, and is by far the most common, though much less simple than that of opposite-leaved buds. How does it correspond with the arrangement of leaf scars on the stem? Refer to Section 52, and say to what order of phyllotaxy it belongs. Notice the gradual change in the size and appearance of the scales from the outside toward the center. Can you give any reasons for regarding them as transformed leaves? Sketch the bud in cross and vertical section (unless this is impracticable on account of the fur) and then remove the contents. Notice the copious fur on the inner scales; of what use is it? Examine with a lens the little furry bodies within the scales and see if you can tell what they are; if you can

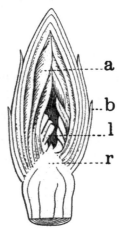

334. — Vertical section of hickory bud; a, furry inner scales; b, outer scales; l, folded leaf; r, receptacle.

not, get a bud that is partly unfolded and you will probably have no trouble in recognizing them as rudimentary leaves.

Notice the manner in which the separate leaflets are folded in the bud and make a diagram of it; how does it differ from that of the buckeye? (Vernation is always best observed in partly expanded buds.) This kind of vernation, in which each leaf or leaflet is rolled over from one side to the other, is called *convolute.* Plum, apple, canna, calla lily, offer good examples of it.

Are there any flower clusters in your hickory bud? if not, look for one that has them. Are they axillary or terminal? Will they stop the further development of their branch? Why or why not?

335. — Expanding bud of English walnut, showing twice conduplicate vernation.

257. Buds with Stipular Scales. — Sketch a bud of the tulip tree, or other magnolia, on the outside. (The India

rubber tree, oak, beech, and hackberry, furnish other examples of stipular scales.) How does it differ in appearance from the ones already examined? Remove the outer pair of scales and observe that (in the tulip tree) their edges do not overlap as in the imbricated arrangement, but merely touch, or in botanical language, are *valvate.* Notice the difference in color between the outer

336. — Bud of tulip tree, showing stipular scales: *s, s,* stipules.

and inner scales. Why are the outer pair so hard and thick? Draw a cross section of the bud as it appears under the lens, showing the small round objects that appear here and there between the scales. Can you make out what they are? Draw a vertical section. Do you see anything like a flower bud? If

337. — Diagram of tulip tree bud in cross section, showing successive leaves (1-7) with stipules (*St.* 5, etc.).

so, is it a cluster or a single flower? (Terminal buds in the tulip tree are usually, but not always, flower buds.) Remove the next pair of scales and notice the rudimentary leaf between them. This outer leaf is often found to be dead; can you account for the fact? Pick off the successive pairs of scales, noticing the leaf between them. Observe that the footstalk of each originates *between* the bases of the scales. You will have no difficulty now in identifying the little round dots in your cross section as the cut ends of the petioles. How many pairs of scales are there in the bud? How many leaflets? Study their arrangement and compare it with the diagram (Fig. 337). How does this correspond with the arrangement of leaves on the stem? Do you find any clusters of bud scale scars as in the other specimens examined?

258. What the Scales are. — The bud scales here clearly can not represent leaves. Compare their position at the foot of the petiole with what was said in Section 32 regard-

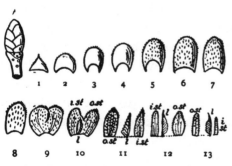

338. — Elm bud with succession of scales: *t,* terminal bud. The scales are numbered in successive order as they occur at the nodes. 9 shows two stipular scales partly fused into one; 10, an outer and an inner stipule, *o. st* and *t. st,* with a rudimentary leaf between; 11, 12, and 13, the same. All are separated to show outline.

ing the stipules, and decide what they are. Notice that the two hard outer ones have no leaflet between them; this is because they are the stipules left by the last leaf of the preceding season, which persist on the stem, though the others usually fall away soon after the leaves develop.

In the elm each scale represents a pair of stipules, as will be evident by observing that they are often notched or bifid at the top, and that the rudimentary leaves stand opposite their scales instead of between them.

259. Arrangement of Scars. — Examine the leaf scars at the nodes of a twig of tulip tree, fig, or magnolia, and notice the ring encircling the stem at each (Fig. 339). These are the scars left by the stipular scales of the past season as they fell away. Where a pair of scales is attached with each separate leaf, they are carried apart as the nodes lengthen, and thus the scars are scattered, a pair at each node all along the stem, instead of being compacted into bands at the base of the bud. They are sometimes very persistent, as in the common fig, where they may often be traced distinctly on stems ten to fifteen years old.

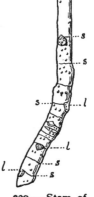

339. — Stem of tulip tree: *s,s*, scars left by stipular scales; *l,l*, leaf scars.

340. — A partly expanded leaf of beech, showing plicate-conduplicate vernation.

260. Vernation. — Notice how the two halves of the leaflets are doubled together by their inner faces and then bent over on the petiole (Fig. 336). The first is called *conduplicate*, and is common in the redbud, rose, peach, cherry, oak, Japan quince, etc.; the second is the *inflexed* mode of vernation. This mixed vernation is very common. In the elm and beech the two halves of the leaf are first plicate and then conduplicate to each other (Fig. 340); in the purple magnolia and chinquapin they are conduplicate-plicate.

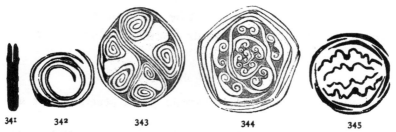

341–345. — Diagrams of vernation: 341, conduplicate (oak); 342, convolute (cherry); 343, revolute (dock); 344, involute (balsam poplar); 345, plicate (sycamore).

261. Forms of Vernation. — The varieties of vernation or prefoliation should be studied and diagrammed as they are met with. In addition to the varieties already mentioned, there are the

Straight: not bent or folded in any way, as Japan honeysuckle, periwinkle, St. John's-wort, dogwood, etc.

Involute (Fig. 344): violet, arrow grass (*Sagittaria*), lotus, water lily, balm of Gilead.

Revolute (Fig. 343): dock, willow oak, scarlet morning-glory (*Ipomea coccinea*), rosemary, azalea, persimmon.

Circinnate (Fig. 346): ferns, sundew.

346. — Circinnate bud of fern.

262. Dormant Buds. — A bud may often lie dormant for months or even years, and then, through the injury or destruction of its stronger rivals, or some other favoring cause, develop into a branch. Such buds are said to be *latent* or *dormant*. The sprouts that often put up from the stumps of felled trees originate from this source.

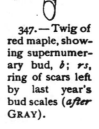

347. — Twig of red maple, showing supernumerary bud, *b*; *rs*, ring of scars left by last year's bud scales (*after* GRAY).

263. Supernumerary Buds. — Where more than one bud develops at a node, as is so often the case in the oak, maple, honey locust, etc., all except the normal one in the axil are *supernumerary* or *accessory*. These must not be confounded with *adventitious* buds, or those that occur elsewhere than at a node.

1. Why do annuals and herbaceous plants generally have unprotected buds? (252.)

2. Why is the gummy coating found on the buds of the horse-chestnut and balm of Gilead wanting in their southern representatives, the buckeye and silver poplar? (252.)

3. Can you name any plants the buds of which serve as food for man?

4. How do flower buds differ in shape from leaf buds?

5. At what season can the leaf bud and the flower bud first be distinguished?

6. Watch any of the trees about your home and see when the buds that are to develop into leaves and flowers the next year are formed.

INFLORESCENCE

MATERIAL. — A few typical flower clusters illustrating the definite and indefinite modes of inflorescence. Some of those mentioned in the text are : —

Indefinite: hyacinth, shepherd's purse, wall flower, parsley, lilac, blue grass, smartweed (*polygonum*), wheat, oak, willow, clover.

Definite: chickweed, spurge (*Euphorbia*, various kinds), comfrey, dead nettle (*Lamium amplexicaule*), etc. Any other examples illustrating the principal kinds of cluster will do as well, but the subject should not be taught without an examination of at least a few living specimens of each sort.

264. Definitions. — Inflorescence is a term used to denote the position and arrangement of flowers on the stem. It is merely a mode of branching and follows the same laws that govern the branching of ordinary stems.

The stalk that bears a flower is called by botanists the *peduncle*. In a cluster the main axis is the common peduncle, or *rhachis*, and the separate flower stalks *pedicels*.

265. Two Kinds of Inflorescence. — The growth of flower stems, like that of leaf stems, is of two principal kinds, definite and indefinite,

348. — Solitary terminal flower of a lily.

or as it is frequently expressed, *determinate* and *indeter-*

minate. The simplest kind of each is the solitary, where a single flower either terminates the main axis, as the daffodil, trillium, magnolia, etc., or springs singly from the axils, as in the running periwinkle, moneywort, and cotton.

349. — Solitary axillary inflorescence of moneywort (*after* GRAY).

266. Indeterminate Inflorescence is always axillary, since the production of a terminal flower would stop further growth in that direction and thus terminate the development of the axis. We have only to imagine the internodes of such a stem or branch as that represented in Figure 349 very much shortened, the leaves reduced to bracts or wanting altogether, and flowers or flower buds at every node, to have the

267. Raceme, the typical flower cluster of the indefinite sort. In such an arrangement the oldest flowers are, necessarily, at the lower nodes, new ones appearing only as the axis lengthens and produces new internodes. This will be made clear by examining a flowering stalk of hyacinth, cherry laurel (*Prunus caroliniana*), shepherd's purse, or any common weeds of the mustard family that are generally to be found in abundance everywhere. It will be seen that the lower buds have already fruited in the last named, and perhaps the pods have dehisced and shed their seed before the upper ones have even begun to unfold. Notice the little scale or bract usually found at the base of the pedicel in

350. — Raceme of milk vetch (*Astragalus*).

flower clusters of this sort (in the shepherd's purse it is wanting). This is a reduced leaf, and the fact that the flower stalk springs from the axil, shows it to be of the essential nature of a branch.

268. The Corymb. — Imagine the lower pedicels of a raceme to be elongated so as to place their flowers on a level with those of the upper nodes, making a convex, or more or less flat-topped cluster, as in the wall-flower and hawthorn, and we have a modification of the raceme called a *corymb.* In such a cluster the outer blossoms, or those on the circumference, proceed from the lower axils and are, consequently, the oldest; hence, the order of flowering is *centripetal,* that is, from the circumference to the center. This, an inspection of Figure 351 will show, is only another way of saying that it is of the indefinite or indeterminate order.

351. — Corymb of plum blossoms.

269. The Umbel is a still further modification of the raceme. The pedicels with their bracts are all gathered

352. — Umbel of milkweed.

at the top of the peduncle, from which they spread in every direction like the rays of an umbrella, as the name implies. This, though confined to no one group, is the prevalent type of flower cluster in the parsley family, which takes its botanical name, *Umbelliferæ,* from its characteristic form of inflorescence. The pedicels of an umbel are generally called *rays* and the circle of bracts at the base of the cluster is an *involucre.*

270. Compound Clusters. — All these forms of inflorescence may be compound. Most of the parsley family have compound umbels. The lilac, grape, catalpa, and many grasses furnish familiar examples of the *panicle,*

353 — Panicle of a grass.

which is merely a compound raceme, the pedicels of which are branched one or more times.

271. A Spike (Fig. 354) is a raceme with the flowers sessile and more or less crowded together, as in the plantain, smartweed, wheat, barley, etc. A form of spike more common in early spring is the

272. Ament, or Catkin, of which we have abundant examples in the pendent scaly inflorescence of the willow, oak, poplar, and most of our common forest trees (Fig. 355). A sessile corymb or umbel gives rise to

354. — A spike of the common plantain (*Plantago lanceolata*).

273. The Head (Fig. 356), a crowded, roundish cluster like the clover, buttonwood, sycamore, etc.

355. — Ament or catkin of birch (*after* GRAY).

356. — Head of clover.

274. Diagrams. — Do not try to learn all these names by heart, but look for examples of the different kinds of inflorescence and diagram them, using balls or circles to symbolize the flowers, as in the models given in Figures 357 to 361. The order of blooming may be shown by using larger balls to represent the older flowers. It will be seen from the diagrams that all the forms of indefinite inflorescence are derived from the raceme, whence it is frequently spoken of as the *racemose* type of inflorescence.

275. Cymose, or Definite Inflorescence. — As the raceme is the fundamental form of indefinite inflorescence, so the fundamental form of the definite or determinate kind is the *cyme*, and hence, the term "cymose" is frequently used as synonymous with determinate or definite.

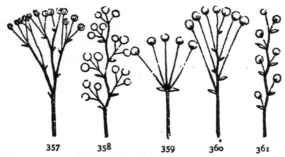

357–361. — Diagrams of indefinite inflorescence: 357, compound corymb; 358, compound raceme, or panicle; 359, umbel; 360, corymb; 361, raceme.

276. Nature of the Cyme. — To understand the nature of the cyme, study a forking branch of common mouse-ear chickweed (*Cerastium vulgatum*), corn cockle, or spurge (*Euphorbia*). Examine carefully what appears to be the topmost cluster of blossoms, and it will be found to consist of a single terminal flower (probably already gone to seed), with two smaller flower clusters rising from the axils of leaves at the base of the peduncle. The older blossoms in the center, being terminal, stopped the growth of the axis in that direction just as we saw in the case of the terminal flower bud of the buckeye, and forced the stem in continuing its growth to send

362. — Forking cyme of common chickweed.

out side branches from the axils of the topmost leaves. One or both of these branches will produce, or perhaps has already produced, in turn, a terminal flower which forces its branch to divide again, and so on, forking indefinitely in a manner precisely analogous to the dichotomous forking of stems like the buckeye and jimson weed. By looking down in the next lower fork you will probably find the remains of a still older flower that terminated the growth in that direction and forced the stem to continue its development by sending off branches on either side, and so on, until the remains of the older flowers have disappeared and the forking becomes obscured. Here the oldest flower is lowest, not because, as in the raceme, the axis has continued to grow beyond it, but because it checked the further development of its own axis and has been overtopped by new branches.

363. — Flat-topped cyme of sneezeweed.

277. Centrifugal Inflorescence. — When the older peduncles are lengthened as described in Section 268, a flat-topped cyme is produced, which is distinguished from the corymb by its *centrifugal* inflorescence ; that is, the oldest flower of each cluster is in the center, and the order of blossoming proceeds from within toward the circumference, as in the star-of-Bethlehem, bitterweed (*Helenium tenuifolium*), etc. If the cyme is much compounded, the inflorescence becomes very complicated, and as many of the blossoms never develop, will seem to have no regular order.

278. The Coiled, or Scorpioid Cyme. — A peculiar form of cyme is found in the coiled inflorescence of the pink-root (*Spigelia*), heliotrope, comfrey, etc. It occurs where a cyme like that represented in Figure 362 develops on one side only. Its structure will be made clear by an inspection of Figures 365–367.

279. Mixed Inflorescence. — We often find the two kinds of inflorescence mixed in the same cluster. In a panicle of buckeye, for example, the whole cluster is terminal with reference to its shoot, while the secondary branches are indefinite, the lower blooming first. The individual flowers

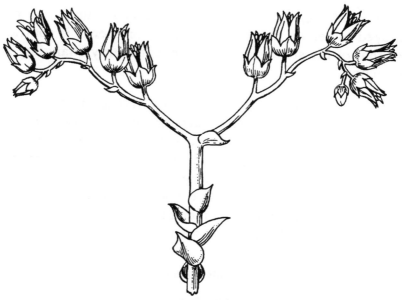

364. — Scorpioid cyme.

of these secondary clusters, again, are of the definite type, being disposed in scorpioid cymes.

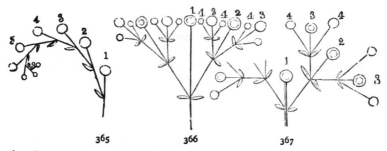

365   366   367

365–367. — Diagrams of cymose inflorescence, with flowers numbered in the order of their development: 365, cyme half developed (scorpioid); 366, a flat-topped or corymbose cyme; 367, development of a typical cyme.

280. Use of Terms. — The distinction between determinate and indeterminate inflorescence is not strictly adhered to in botanical descriptions, especially if the clusters are at all complicated. It is well to remember, however, that the terms indefinite, indeterminate, racemose, centripetal, all mean about the same thing; namely, that the flowers develop with the axis, or from below upward; and the terms definite, determinate, cymose, centrifugal, are employed to denote that the order of inflorescence is contrary to that of the stem growth, and is constantly changing its direction.

281. Significance of the Clustered Arrangement. — As a general thing the clustered arrangement marks a higher stage of development than the solitary, just as in human life the rudest social state is a distinct advance upon the isolated condition of the savage. In plant life it is the beginning of a system of coöperation and division of labor among the associated members of the flower cluster, as will be seen later, when we take up the study of the flower.

PRACTICAL QUESTIONS

1. Name as many solitary flowers as you can think of.
2. Do you find very small flowers, as a rule, solitary, or in clusters?
3. Would the separate flowers of the clover, parsley, or grape, be readily distinguished by the eye from among a mass of foliage?
4. Should you judge from these facts that it is, in general, advantageous to plants for their flowers to be conspicuous?

FIELD WORK

The foregoing lessons are themselves so full of suggestions for field work that it hardly seems necessary to add anything to them.

In connection with Sections 240–248, the characteristic modes of branching of the common trees and shrubs of each neighborhood should be observed and accounted for. The naked branches of the winter woods afford exceptional advantages for studies of this kind, which can not well be carried on except out of doors. Trees should be selected for observation that have not been pruned or tampered with by man. Note the effect of the mode of branching upon the general outline of the tree; compare the direction and mode of growth of the larger boughs with that of small twigs in the same species and see if there is any general correspondence between them; note the absence of fine spray on the boughs of large-leaved trees, and account for it. Account for the flat sprays of trees like the elm, beech, hackberry, etc.;

the irregular stumpy branches of the oak and walnut; the stiff, straight twigs of the ash; the zigzag switches of the black locust, Osage orange, elm, linden, etc. Measure the twigs on various species and see if there is any relation between the length and thickness of branches. Notice the different trend of the upper, middle, and lower boughs in most trees and account for it. Observe the mode of branching of as many different species as possible of some of the great botanical groups of trees; the oaks, hickories, hawthorns, or pines, for instance, and notice whether it is, as a general thing, uniform among the species of the same group, and how it differs from that of other groups.

In connection with Sections 249–263, buds of as many different kinds as possible should be examined with reference to their means of protection, their vernation and phyllotaxy, and the modes of growth resulting from them. Compare the folding of the cotyledons in the seed with the vernation of the same plants, and observe whether the folding is the same throughout a whole group of related plants, or only for the same species. Notice which modes seem to be most prevalent. Select a twig on some tree near your home or your schoolhouse and keep a record of its daily growth from the first sign of the unfolding of its principal bud to the full development of all its leaves. Any study of buds should include an observation of them in all stages of development.

With Sections 264–281, study the inflorescence of the common plants and weeds that happen to be in season, until you have no difficulty in distinguishing between the definite and indefinite sorts, and can refer any ordinary cluster to its proper form. Notice whether there is any tendency to uniformity in the mode of inflorescence among flowers of the same family. Consider how each kind is adapted to the shape and habit of the flowers composing it, and what particular advantage each of the specimens examined derives from the way its flowers are clustered. In cases of mixed inflorescence see if you can discover any reason for the change from one form to the other.

VIII. THE FLOWER

HYPOGYNOUS MONOCOTYLEDONS

MATERIAL. — Any flower of the lily family with disunited petals. Star-of-Bethlehem and yucca are used in the text. Tulip, trillium, dogtooth violet (*Erythronium*), spiderwort (*Tradescantia*), white lily, all make excellent examples.

282. The Floral Envelopes. — Make a sketch of a flower of the star-of-Bethlehem, or other of the lily tribe, from the outside. Label the head of the peduncle that supports the flower, *receptacle*, or *torus*, the three outer greenish leaves, *sepals*, the three inner, lighter colored ones, *petals*. The sepals taken together form the *calyx*,

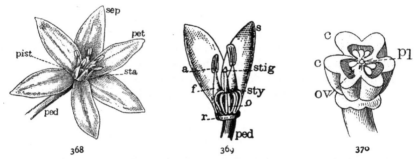

368 369 370

368–370. — Flower of a hypogynous monocotyledon dissected: 368, a flower of the star-of-Bethlehem, showing the different sets of organs: *pet*, petals; *sep*, sepals; *sta*, stamens; *pist*, pistil; *ped*, peduncle; 369, side view of star-of-Bethlehem with all the petals and sepals but two removed to show order of the parts: *r*, receptacle; *o*, ovary; *sty*, style; *stig*, stigma — parts composing the pistil; *f*, filament; *a*, anther — parts composing the stamen; 370, cross section of the ovary of star-of-Bethlehem: *c, c*, carpels; *ov*, ovules; *pl*, placenta.

and the petals, the *corolla*. In many flowers, such as the tulip and Atamasco lily (*Zephyranthes*), there is little or no difference between them. In such cases the calyx and corolla together are called the *perianth*, but the distinction of parts is always observed, the three outer divisions being regarded as sepals, the inner ones as petals. These two sets of organs constitute the *floral envelopes*, and are not essential parts of the flower, as it can fulfill its office of producing fruit and seed without them. Note their mode of attachment to the receptacle and how they alternate with each other. Remove one of the sepals and one of the petals, and notice any

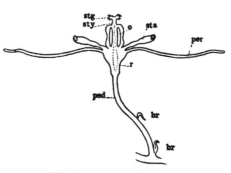

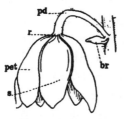

371. — External view of a yucca blossom : *br*, bract ; *pd*, peduncle ; *r*, receptacle ; *s*, sepal ; *pet*, petal.

372. — Vertical section of yucca whipplei : *ped*, peduncle ; *br*, bract ; *r*, receptacle ; *per*, perianth ; *sta*, stamen ; *o*, ovary ; *sty*, style ; *stg*, stigma. The last three parts named compose the pistil.

differences between them as to size, shape, or color. Which is most like a foliage leaf? Hold each up to the light and try to make out the veining. Is it the same as that of the foliage leaves? How many of each are there?

283. The Essential Organs. — Next sketch the flower on its inner face, labeling the six appendages just within the petals, *stamens*, and the central organ within the ring of stamens, *pistil*. These are called essential *organs* because they are necessary to the production of fruit and seed. Note their mode of insertion, three of the stamens alternating with the petals and the other three with these, and with the lobes of the base of the pistil.

284. The Stamens. — Notice whether the stamens are all alike, or whether there are differences as to size, height, shape, color, etc. Do these differences, if there are any, occur indiscriminately and without order, or in regular succession between the alternating stamens? Examine one of the little powdery yellow bodies at the tip of the stamens, and see whether they face toward the pistil or away from it. In the first case they are said to be *introrse*, in the second, *extrorse*.

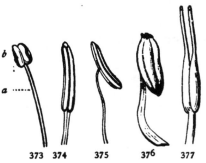

Observe the mode of attachment of the anthers, whether by their base merely (*terminal*), or through their entire length (*adnate*), or to the tip of the filament as on a pivot, so as to admit of their turning freely in all directions (*versatile*).

373–377.—Stamens (GRAY): 373, a stamen with the anther, *b*, surmounting the filament, *a* (terminal), and opening in the normal manner down the outer side of each cell; 374, stamen of tulip tree, with adnate extrorse anther; 375, stamen of an evening primrose (*Œnothera*) with versatile anther; 376, stamen of pyrola, the anther cells opening by chinks or pores at the top; 377, stamen of a cranberry, with the anther cells prolonged into a tube and opening by a pore at the apex.

Remove one of the stamens and sketch it as it appears under the lens, labeling the powdery yellow body at the top, *anther*, the stalklike (in the star-of-Bethlehem expanded and petal-like) body supporting it, *filament*. Usually the filaments are threadlike, whence their name, but in the star-of-Bethlehem they look like altered petals, and frequently a stamen is found in a transition state, as if changing from stamen to petal, or from petal back to stamen. See if you can find such a one. What would you infer from this fact?

Notice the two little sacs or pouches that compose the anther, as to their shape and manner of opening, or dehiscing, to discharge the powder contained in them. This powder is called *pollen*, and will be seen under the lens to consist of little yellow grains. These are of

378–381.—Forms of pollen (GRAY): 378, from *mimulus moschatus;* 379, *sicyos;* 380, *echinocystis;* 381, *hibiscus.*

different shapes, colors, and sizes, in different plants, and the surface is often beautifully grooved and striate. The grains with their markings are always alike in the same species, so that it is possible to recognize a plant by its pollen alone. These characters are generally too minute to be observed without a compound microscope, but in the hibiscus, and some others of the mallow family, they can be distinguished with a hand lens.

285. The Pistil. — Remove the stamens and sketch the pistil as it stands on the receptacle. Label the round or oval enlargement at the base, *ovary*, the threadlike appendage rising from its center, *style*, and the tip end of the style, *stigma*. If the stigma is lobed or parted, count the divisions and see if there is any correspondence between them and the number of petals and sepals, or of the lobes of the ovary. Examine the tip with a lens and notice the sticky, mucilaginous exudation that moistens it. Can you think of any use for this? If not, touch one of the powdery anthers to it, and examine it again with the lens. What do you see?

286. Pollination, or the transfer of pollen from the anther to the stigma, is a matter of great importance, as the pistil can not develop seed without it. Note the relative position of pistils and stamens and see if it is such that the pollen can reach the stigma without external agency.

287. The Ovary. — Observe the shape of the ovary, and the number of ridges, or grooves that divide the surface. These lines correspond to the sutures of the fruit, and show of how many carpels the ovary is composed. In the star-of-Bethlehem the ovary has six sutures, three of which represent the midrib of the carpellary leaves, and three

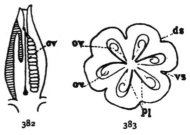

382, 383. — Ovary of *yucca aloifolia*, a hypogynous monocotyledon, dissected: 382, vertical section: *ov*, ovules; 383, diagram of a horizontal section of the same, enlarged, showing the three carpels and six cells, or loculi: *ds*, dorsal sutures; *vs*, ventral sutures; *ov*, ovules; *pl*, placenta.

the inner or ventral sutures, so that there are only three true carpels. Select a flower that has begun to wither, so that the ovary is well developed, cut a cross section near the middle and try to make out the number of cells, or internal divisions. Make an enlarged sketch of the section as it appears under the lens (see Fig. 383), showing the arrangement of the parts, also a longitudinal section (Fig. 382) showing their relative vertical position. Label the little round bodies that represent the undeveloped seeds *ovules*, the surface to which they are attached, *placenta*, and the cavities, or divisions containing them, *cells*, or *loculi* (singular, *loculus*). How many of these are there? Compare these sketches of the ovary with your drawings of dehiscent fruits in Sections 93–109. What correspondences do you notice between them?

As the ovary is merely an undeveloped fruit, and the ovules immature seeds, their structure is the same as that of these parts, and the same terms are used in describing them (Secs. 73–79, and 93–109).

288. Numerical Plan.—Now make a horizontal diagram, after the model given in Figure 384, showing the manner of

384.— Horizontal diagram of a flower of the lily kind. The dot represents the growing axis of the plant.

attachment of the different cycles — sepals, petals, stamens, and pistils, the number of organs in each set, and their mode of alternation with the organs of the other cycles. Notice that in the star-of-Bethlehem and similar flowers, the parts of each set are in threes, or multiples of three. This is called the numerical plan of the flower, and is the prevailing number among monocotyledons. It is expressed in botanical language by saying that the flower is *trimerous*, a word meaning measured, or divided off into parts of three.

289. Vertical Order. — Next make a vertical diagram of your specimen after the manner shown in Figure 372, and note carefully that the ovary stands *above* the other organs (this is true of all the lily family), and is entirely separate and distinct from them. In such cases the ovary is said to be *free*, or *superior*, and the other organs *inferior*, or *hypogynous*, a word meaning "inserted under the pistil." These terms should be remembered, as the distinction is . an important one in plant evolution.

290. The Flower Bud. — Observe the manner in which the sepals and petals overlap in a partly unfolded bud. Draw a diagram representing their position, as in Figures 385–387. Compare this with your diagrams of leaves and leaf buds; does it agree with any of them, and if so, which? Are the parts imbricated or valvate? (Secs. 250, 256, 257.)

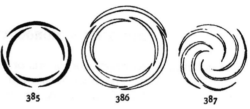

385–387. — Diagrams of three modes of æstivation common among monocotyledons: 385, valvate; 386, imbricate (GRAY); 387, convolute (GRAY).

The arrangement of the parts of the flower in the bud is called *æstivation*, or *prefloration*, words meaning respectively "summer condition" and "condition before flowering." It corresponds to the vernation of leaf buds, and the same terms are used in describing it.

291. Summary of Observations. —In the flower just examined we found that there were four sets of floral organs present — sepals, petals, stamens, and pistil; that the individual organs in each set were alike in size and shape; that there were the same number, or multiples of the same number of parts in each set, and that all the parts of each set were entirely separate and disconnected the one from the other, and from those of the other cycles. Such a flower is said to be : —

Perfect, that is, provided with both kinds of organs essential to the production of seed — stamens, and pistil.

Complete, having all the kinds of organs that a flower can have; viz.: two sets of essential organs, and two sets of floral envelopes.

Regular, having all the parts of each set of the same size and shape.

Symmetrical, having the same number of organs, or multiples of the same number in each set,

The opposites of these terms are: imperfect, incomplete, irregular, and asymmetrical, or unsymmetrical.

Note that regularity refers to form, symmetry to number of parts, and that a flower may be perfect without being complete.

EPIGYNOUS MONOCOTYLEDONS

MATERIAL. — Any flower of the iris or amaryllis families. Iris is used in the text. Blackberry lily (*Belamcanda*), Atamasco lily (*Zephyranthes*), snowdrop, daffodil, narcissus, etc., will make good examples.

292. The Perianth. — Compare with the flower last examined, a common flag, or iris. Notice that the latter has no peduncle, but is sessile in the axil of a large, membranous bract called a *spathe.* Observe also that the lower part of the perianth is united into a long,

388. — Iris flower: *sp*, spathes; *s*, sepals + *p*, petals = perianth.

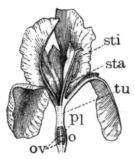

389. — Vertical section of iris flower (*after* GRAY): *ov*, ovules; *pl*, placenta; *tu*, tube of the perianth inclosing the style; *sta*, stamen; *sti*, stigma.

narrow tube, from the top of which the sepals and petals extend as long, curving lobes. Where the parts of a perianth or of a corolla are united in this way, whether throughout their whole length, as in the morning-glory, or by a mere thread or rim at the base, as in the water pimpernel, it is said to be *sympetalous*, meaning "of united petals." Monopetalous and gamopetalous are other words used to denote the same thing, and the kindred terms, synsepalous, gamosepalous, etc., are applied to the calyx.

293. Dissection of the Iris. — Sketch the outside of the specimen, labeling the oblong, three-lobed enlargement at the base, *ovary*, the prolongation of the flower above it, *tube of the perianth*, the three outer lobes with the broad sessile bases, *sepals*, the others, with their bases narrowed and bent inward, *petals*. Now turn the flower over and sketch the inside, labeling the three large, petal-like expansions in the center, *stigmas*. Do you see any stamens? Remove one of the sepals and look under the stigma; what do you find there? Notice the little honey pockets at the foot of the stamen. Run the head of your pencil into them and see what would happen to the head of an insect probing for honey.

Remove all the petals and sepals and sketch the remaining organs in profile, showing the position of the stamens. Are the anthers extrorse or introrse? What is their mode of dehiscence? Remove a stamen and sketch it. What is the shape of the anther?

390. — Vertical section of iris flower, with perianth removed, showing a stamen and three stigmas: *su*, stigmatic surface.

Remove as much of the upper part of the perianth tube as you can without injuring the pistil, and with a sharp knife, slice away a section down through the ovary so as to show the long style and its connection with the placenta. Make a sketch of this longitudinal section (see Fig. 389), labeling the long, club-shaped stalk running from the ovary to the stigmas, *style*; the white column in the center of the ovary to which the undeveloped seed are attached, *placenta*, and the unripe seeds, ovules. Notice whether the placenta is central or parietal (Secs. 103, 109). Draw a cross section of the ovary; how many cells has it? Examine with a lens the little flap under the two-cleft apex of one of the stigmas, and

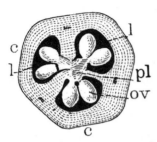

look for a moist spot to which the pollen will adhere. Label this in your longitudinal sketch, *stigmatic surface*. No seeds can be matured unless some of the pollen reaches this surface; can you think by what agency it is carried there? What insects have you seen hovering about the iris? Notice that in draw-ing his head *out* of the flower, an insect would not touch the stig-

391. — Cross section of ovary of iris flower: *c, c,* car-pels; *l, l,* cells, or loculi; *ov,* ovules; *pl,* placenta.

matic surface, since it is on the *upper* side of the flap and he would be probing *under* it. But in entering the next flower that he visits, he is likely to strike his head against the flap and turn it under, thus dusting it with pollen brought from another flower.

Sketch a sepal and a petal separately, and note their differences as to shape, color, and texture. Hold each up to the light and observe the veining. If this is not clear, stand a specimen in red ink for two or three hours and examine it again. Is it parallel or net veined? Can you think of a use for the crest of hairlike filaments on the upper side of the sepals?

Examine a bud in cross section. Notice how the sepals and petals overlap, and draw a diagram of the section. This manner of arrangement, where the outer edge of one piece covers the inner edge of the one next above it (Fig. 387), is said to be *convolute*. Draw diagrams showing the horizontal and vertical arrangement of parts in the iris. What is its numerical plan? Is it symmetrical? Regular? Are the parts all free? If not, which are united among themselves or with other sets of organs? Is the ovary above or below the other parts?

392. — Horizon-tal diagram of iris flower.

294. The Epigynous Arrangement. — In cases of this kind, where the other organs appear to rise from the top of the ovary, they are said to be *epigynous*, a word meaning "upon the ovary." The same thing is expressed in a different way by saying that the ovary is *inferior*, or that the other organs are *superior*. To make the matter clear, the two sets of terms employed for describing the position of the ovary are given below in parallel columns.

Hypogynous	Epigynous
Ovary superior	Ovary inferior
Calyx or perianth inferior	Calyx or perianth superior

The epigynous arrangement is considered to mark a higher stage of floral development than the hypogynous, which is characteristic of a more simple and primitive structure.

DICOTYLEDONS

MATERIAL. — Blossoms of any convenient specimens of the mustard family. Large flowered species are always best if they can be obtained; cabbage, mustard, turnip, and wall-flower are very good.

Flowers of apple, pear, or quince, and of peach, plum, cherry, or rose; also of any member of the pea family, such as bean, pea, vetch, black locust, wistaria, etc.

295. Dissection of a Typical Flower. — Gently remove the sepals and petals from a mustard or other cress flower, lay them on the table before you in exactly the order in which they grew on the stem, and sketch them. How many of each are there, and how do they alternate with one another? Sketch the pistil and stamens as they stand on the receptacle; how many of the latter are there? Notice that two of the six are outside and a little below the others, alternate with the petals, while the other four stand opposite them, as is natural if they were alternating with another ring of stamens between themselves and the corolla. Stamens arranged in this way are said to be *tetradynamous*, that is, four stronger, or larger than the others. Put a dot before two of the sepals in your first drawing to indicate the position of the two outer stamens, and a cross before the other two to show where stamens are wanting to complete the symmetry of this set as in the diagram (Fig. 395). When parts necessary to complete the plan of a flower are wanting, as in this case, they are said to be *obsolete, suppressed,* or *aborted.* Place dots before the petals to represent the other four stamens.

Examine the anthers under the lens. Are they extrorse or introrse? What is their mode of attachment to the filament? (Sec. 284.) Sketch one of the anthers, show-

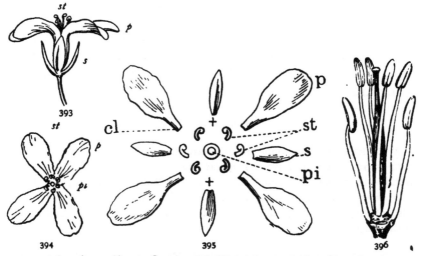

393-396. — A cruciferous flower: 393, side view. 394, view from above. 395, diagram of parts: *p*, petals; *s*, sepals; *st*, stamens; *pi*, pistil; *cl*, claw of petal; +, +, position of the missing stamens. 396, pistil and stamens, enlarged (GRAY).

ing the sagittate base. Remove all the stamens and sketch the pistil, showing the long, slender ovary, the very short style, and the capitate (round and knoblike) stigma. Compare the pistil with a more matured one from an older flower lower down on the stem, and with the descriptions of dehiscent fruits in Sections 93–109, and decide to which kind it belongs. Represent its position by a small circle in the center of your sketch of the separate parts. You have now a complete ground plan of the flower. To what form of leaf arrangement does it correspond? Diagram a vertical section showing the position of the ovary with reference to the other parts, and report in your notebook as to the following points: —

Numerical plan	Presence or absence of parts
Symmetry	Union of parts
Regularity	Position of ovary

A flower put up on the plan of four, like the one just examined, is said to be *tetramerous*, or four parted. The cress or mustard family gets its botanical name, *Cruciferæ*, cross-bearers, from the four opposite petals, which have somewhat the appearance, when viewed from above, of a St. Andrew's Cross. The cruciferous flowers and tetradynamous stamens are striking characteristics of this family, which is so well marked that the merest beginner can hardly fail to recognize any member of it. Notice that its flowers belong to the hypogynous class.

296. Dissection of an Epigynous Dicotyledon. — Sketch a blossom of quince, haw, pear, or apple, first from the outside, then from the inside, and then in vertical section, labeling the parts as in your other sketches. Notice how the ovary is sunk in the hollowed-out receptacle (Sections 74, 77). Where are the other parts attached? Are they inferior or superior? Hold up a petal to the light and examine its venation through a lens. (Use for this purpose a petal from a flower that has stood in red ink for two or three hours. The cherokee rose petals show venation beautifully.) Is it parallel veined or net veined?

397–400. — Flower and sections of pear: 397, cluster of blossoms, showing inflorescence; 398, vertical section of a flower; 399, ground plan of a flower; 400, vertical section of fruit.

Remove a stamen and sketch it as it appears under the lens. Notice the attachment and shape of the anthers. Are they all of the same color? How do you account for the difference, if there is any? Is the position of the pistil and stamens such that the pollen from the anthers can readily reach the stigmas without external aid? Examine the pistil in flowers of different ages, and see if the stigma is mature (that is, moist and sticky) at the same time that the anthers are discharging their pollen.

Draw a cross section of the ovary and try to make out with a lens the number of cells, or loculi. If you can not succeed, turn to the cross section of the pome made in your study of fruits, and that will settle the question, since the fruit is merely a ripened ovary.

401–403. — Types of imbricated æstivation common among dicotyledons (*after* GRAY).

Examine the overlapping of the petals in the bud, and diagram their æstivation (Figs. 401–403).

Compare this with the diagrams of leaf arrangement in Sections 50–52, and decide to which it corresponds.

Diagram the plan of the flower in cross and vertical section. How many parts are there in each set? Can you readily tell the number of stamens? When the individuals of any set or cycle of organs are too numerous to be easily counted, like the stamens of the apple, pear, and peach, or the petals of the water lily, they are said to be *indefinite*. It is very seldom that perfect symmetry is found in all parts of the flower. The stamens and pistil, in particular, show a great tendency to variation, so that the numerical plan is generally determined by the calyx and corolla. Where the parts are in fives, as in the pear, quince, wild rose, etc., the flower is said to be *pentamerous*, or in sets of five.

After drawing the diagrams, write in your notebook answers to the following questions : —

What is the numerical plan of the flower?
Which of its circles of organs is lacking in symmetry?
Which sets of organs are adherent to other sets?
Is the flower epigynous or hypogynous?

297. Examination of a Perigynous Flower.—Compare with the specimen just examined, a blossom of peach, almond, plum, or cherry. Is its numerical plan the same? Make a diagram showing the arrangement of parts in vertical section. Is the calyx inferior or superior? Where are the petals and stamens inserted?

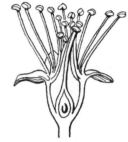

404.—Vertical section of an almond blossom with petals removed, showing the perigynous arrangement.

Flowers of this kind, where the ovary is free and the other parts attached to a prolongation of the receptacle containing it, are said to be *perigynous*, meaning "around the pistil." It is intermediate between the hypogynous and epigynous arrangement, sometimes approaching more nearly to the latter, as in the rose, sometimes remaining clearly of the hypogynous type, as

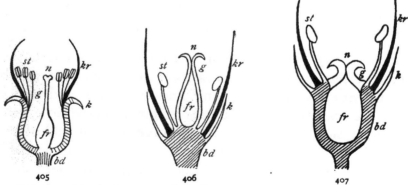

405   406   407

405-407.—Diagrams showing arrangement of parts (*bd*, receptacle; *k*, calyx; *kr*, corolla; *st*, stamens; *fr*, ovary; *g*, style; *n*, stigma): 405, perigynous; 406, hypogynous; 407, epigynous.

in the peach and cherry. In general a flower is not considered epigynous unless the ovary is more or less consolidated with the parts around it.

298. Dissection of an Irregular Flower. — Irregularity is more noticeable in the corolla than in the other parts, and when we speak of an irregular flower the reference is generally to that organ.

Sketch a blossom of any kind of pea or vetch as it appears on the outside. Are the sepals all of the same length and shape? If not, which are the shorter, the upper or lower?

Turn the flower over and examine its inner face. Notice the large, round, and usually upright petal at the back, the two smaller ones on each side, and the boat-

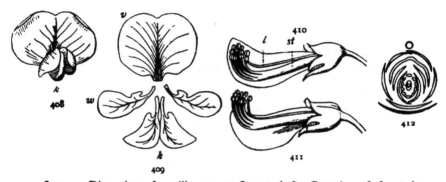

408–412. — Dissection of papilionaceous flowers (*after* GRAY): 408, front view of a corolla. 409, the petals displayed: *v*, vexillum, or standard; *w*, wings; *k*, keel. 410, side view with all except one of the lower petals removed, showing the essential organs protected in the keel: *l*, loose stamen; *st*, stamen tube. 411, side view, showing how the anthers protrude when the keel is depressed. 412, ground plan.

shaped body between them, formed of two small petals more or less united at the apex. Press the side petals gently down with the thumb and forefinger and notice how the essential organs are forced out from the little boat in which they are concealed. Observe how the end of the style is bent over so as to bring the stigma uppermost when the petals are depressed. Imagine the legs of a bee or a butterfly probing for honey; with what organ would his body first come in contact when he alighted? If his thorax and abdomen had previously become dusted with pollen when visiting another flower, where would the pollen be likely to be deposited?

Remove the sepals and petals from one side and sketch

the flower in longitudinal section, showing the position of the pistil and stamens. Then remove all the petals, and spread in their natural order on the table before you, and sketch as they lie (Fig. 409). Label the large, round upper one, *vexillum*, the smaller pair on each side, *wings*, and the two more or less coherent ones in which the pistil and stamens are contained, *keel*. Corollas of this kind are named *papilionaceous*, from the Latin word *papilio*, a butterfly, on account of their general resemblance to that insect; while the old names are somewhat incongruous, they are descriptive, and answer their purpose sufficiently well to be retained.

299. Dissection (*continued*). — Count the stamens, and notice how they are united into two sets of nine and one. Stamens united in this way, no matter what the number in each set, are said to be *diadelphous*, that is, in two brotherhoods. Notice the position of the lone brother, whether below the pistil — next to the keel — or above, facing the vexillum. Would the projection of the pistil when the wings are depressed be facilitated to the same extent if the opening in the stamen tube were on the other side, or if the filaments were *monadelphous* — all united into one set? Flatten out the stamen tube, or sheath formed by the united filaments, and sketch it.

Remove all the parts from around the pistil, and sketch it as it stands upon the receptacle. Look through your lens for the stigmatic surface (Sec. 293). See if there are any hairs upon the style, and if so, whether they are on the front, the back, or all around. Can you think of a use for these hairs?

300. Dissection (*continued*). — Notice how the long, narrow ovary is attached to the receptacle; is it sessile, or raised on a short footstalk? If the latter, label the footstalk *stipel*. Select a well-developed pistil from one of the lower flowers, open the ovary parallel with its flattened sides and sketch the two halves as they appear under the lens. Notice to which side the ovules are attached, the upper (toward the vexillum) or the lower, and label it

placenta. Which suture of the pod is this (Sec. 98)?
Compare with your sketches of dehiscent
fruits; which one does it resemble?

Examine a bud and diagram the æstivation. Which petal overlaps the others?
Diagram the flower in horizontal and vertical section, and decide upon the following
points : —

413. — Diagram
of æstivation of a
papilionaceous
corolla.

What is the numerical plan?

In what organ or organs is there a departure from
symmetry?

In which is there irregularity?

Are all the parts free?

In which set of organs is there union?

Is the flower hypogynous or epigynous?

301. Significance of these Distinctions. — These distinctions are important to remember not only because they are
very useful in grouping and classifying plants, but because
they mark successive stages in the evolution of the flower.
In general, flowers of a primitive type and less advanced
organization are characterized by having their organs free
and hypogynous, while the more highly developed forms
show a tendency to consolidation and union of parts, and
the epigynous mode of insertion. Irregularity also, since
it indicates specialization and adaptation to a particular
purpose, may be regarded as a mark of advanced evolution.

302. Numerical Plan of Dicotyledons. — In all the flowers
examined in Sections 295–300 except the first specimen,
the organs were found to be in fives, or multiples of five.
This is the prevailing number among dicotyledons, though
other orders are not uncommon, and occasionally even
trimerous forms like the magnolia, pawpaw, etc., are met
with. In the mustard family, in the common yellow
primroses of our old fields, and in several other well-
known species, the tetramerous, or fourfold arrangement
prevails, while some of the saxifrages, and a few other
plants are *dimerous*, having their parts in twos. For the

sake of brevity these terms are generally written, in botanical descriptions, 2merous, 3merous, 4merous, 5merous, which are pronounced respectively, dimerous, trimerous, etc.

THE COROLLA

MATERIAL. — Practical illustrations of Sections 303–318 must be sought for out of doors, by observing the various flowers and weeds with which the student comes in contact in his daily walks.

303. Cohesion and Adhesion. — A flower that is perfectly symmetrical and regular, with all its parts free and distinct, like the star-of-Bethlehem and most of the lily family, is not often met with. Frequently one or more of the organs are wanting; more frequently still they are combined and consolidated in various ways with each other or with organs of a different set. Union between organs of the same set is called *cohesion;* between organs of different kinds, *adhesion,* or *adnation.* The opposite of coherent is *distinct;* of adherent, *free.*

304. Apopetalous and Sympetalous Corollas. — Consolidation may occur between any parts of the flower, either of the same or of different sets, but is more conspicuous in the corolla, so that this character has been made the basis of one of the great divisions of seed-bearing plants, which are classed as *apopetalous* and *sympetalous,* according as their corollas are composed of separate or of united petals. Flowers that have no corolla are

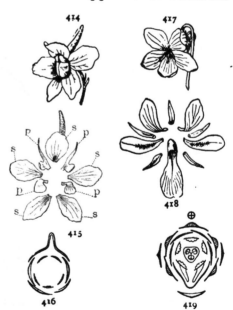

414–419. — Irregular apopetalous corollas (*after* GRAY): 414, a larkspur flower; 415, sepals, *s, s,* and petals, *p, p,* displayed; 416, diagram of arrangement; 417, corolla of the violet; 418, sepals and petals displayed; 419, diagram of arrangement.

said to be *apetalous*, that is, without petals. The term *polypetalous* is sometimes used instead of apopetalous.

305. Apopetalous Corollas may have any number of petals, from one or two, as in the enchanter's nightshade (*Circæa*), to the indefinite whorls of such double flowers as the cactus and water lily. They may be of all shapes and sizes, and sometimes present the greatest irregularities of structure, as the violet, tropæolum, larkspur, and columbine. The commonest type of irregular corolla belonging to the apopetalous group, and the only one that has received a special name, is the papilionaceous corolla already described, that characterizes the pea family. This may well be called the reigning family of this division, since it is by far the most important and numerous, containing about seven thousand known species, among which are many of our most useful food plants.

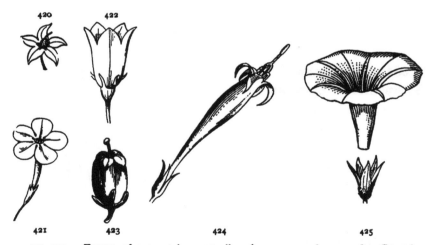

420-425. — Forms of sympetalous corollas (420-422, and 425, *after* GRAY): 420, rotate corolla of nightshade; 421, salver-shaped corolla of phlox; 422, campanulate corolla of harebell; 423, urceolate, or urn-shaped corolla of andromeda; 424, tubular corolla of spigelia; 425, funnel-shaped corolla of morning-glory.

306. Sympetalous Corollas are of so many different forms that it has been found convenient to apply special names to the more important of them. A correct idea of these can be gained by comparing living specimens as they are found with Figures 420-425.

307. The Ligulate, or strap-shaped corolla, seen in the rays of the sunflower family, is of such frequent occurrence as to deserve a special examination. If you will remove one of the small blossoms from the disk of any large composite flower (Fig. 426) and imagine its corolla greatly enlarged and split open on the inner side, you will get a very good idea of the nature of the

426. — A head of artichoke flower divided lengthwise.

427. — A ray flower of artichoke, enlarged.

rays. The five little teeth into which it is usually cleft at the top show the number of lobes or petals of which it is composed. The corolla of the lobelia represents an intermediate state between the tubular and ligulate forms (Fig. 429).

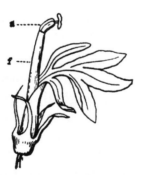

428. — A vertical section of a disk flower, showing the divided style, *st*, and the stamens, *s, s*, with their anthers united (syngenesious).

429. — Flower of *Lobelia cardinalis*, with tube of corolla divided on one side; filaments and anthers united into a tube (*after* GRAY): *f*, tube of filaments; *a*, anthers.

308. Bilabiate Corollas. — By. far the most important and widely distributed of sympetalous corollas is the bilabiate, or two-lipped kind, distinctive of the mint and figwort families and their allied groups, numbering in all over six thousand known species. They are of many varieties, from the scarcely perceptible irregularity of the verbena and mullein to the complicated structures of the sage, snapdragon, and toad flax. Two of them are so strongly marked that they have received special names. These are the *ringent*, or open-mouthed, and the *personate*, or closed (Figs. 430

and 431), so called from a fancied resemblance of the swollen palate to a grotesque *persona*, or mask. The sage and dead nettle are familiar examples of the first, the snapdragon and toadflax of the second. An inspection of the sage or the dead nettle will show that the two lips represent the divisions of a five-lobed sympetalous corolla united into sets of two and three petals respectively. The very divergent appendage of the lower lip represents the middle one of three petals, while the two lateral ones have become greatly reduced, or in the dead nettle, nearly

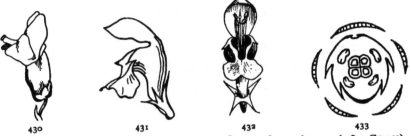

430   431   432   433

430–433. — Bilabiate corollas: 430, personate flower of snapdragon (*after* GRAY) ; 431, ringent corolla of dead nettle; 432, front view; 433, horizontal diagram.

obsolete. The arched upper lip represents two petals confluent into one, a notch in many species (catnip, dittany, snapdragon), indicating the original line of division.

Some of the names given to sympetalous corollas apply equally to apopetalous ones. Chickweed and moonseed are rotate; the uvularias, the yucca, and the abutilon of the greenhouses are bell-shaped, or campanulate; okra and some of the lilies are funnel-shaped.

The same terms that are used in describing the shapes of foliage leaves are applied to the sepals and petals of flowers.

SUPPRESSIONS, ALTERATIONS, AND APPENDAGES

MATERIAL is to be sought for out of doors, wherever it may present itself. Specimens of pine, oak, or other unisexual flowers should be provided for class study. If these are not in season, the mulberry, Osage orange, hop, sycamore, black gum, persimmon, and the gourds, squashes, and melons, furnish good examples of unisexual flowers, one or more of which ought to be examined.

309. Undeveloped Organs. — A flower may depart from
the normal type either by the non-development of parts,
or through the suppression or alteration of
parts already developed. A want of development generally characterizes simple and
primitive forms such as the naked flowers
of the lizard's tail (*Saururus*), the black ash,

and willow, in which the
floral envelopes are entirely
lacking, or reduced to a
mere scale or bract. A step higher in

434. — Naked
flower of *Sau-
rurus* (*after
GRAY*).

435. — Petal-like sepals
of clematis.

the order of development the floral envelopes appear, but are usually inconspicuous and without differentiation into
calyx and corolla, as in the elm, knot-
weeds, docks, etc. Where only one set of these organs is
present, it is considered a calyx, no matter how large and
conspicuous it may be, as in the four-o'clock, and clematis.

310. Unisexual Flowers. — Where one of the essential
organs is lacking, the flower is *unisexual*, which means that
either stamens only, or pistils only, occur in the same
flower. When the stamens alone are
present the flower is said to be staminate, or *sterile* because it is incapable
of producing seeds of its own, though
its pollen is a necessary factor in their
production. If, on the other hand, the
ovary is present and the stamens
absent, the flower is pistillate and *fertile;* that is, capable of producing fruit
when impregnated with pollen. Sometimes both stamens and pistils are
wanting, as in the showy corollas of

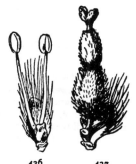

436   437

436, 437. — Flowers of
willow : 436, pistillate ;
437, staminate.

the garden "snowball" and hydrangea, and the rays
of the sunflower. Such blossoms are said to be *neutral*,
from the Latin word *neuter*, neither, because they have
neither pistils nor stamens. They can, of course, have no

direct part in the production of fruit, but are for show merely. Their show, however, is far from being a vain and empty one, as we shall see in Sections 330–338.

311. Monœcious and Diœcious Plants. — When both kinds of flowers, staminate and pistillate, are borne on the same plant, as in the oak, pine, hickory, and most of our common forest trees, they are said to be *monœcious*, a word which means "belonging to one household," and *diœcious*, or "of two households," when borne on separate plants, as in the willow, sassafras, and black gum. Draw a flowering twig of oak, or other amentaceous (ament-bearing) tree. Where are the fertile flowers situated? Notice how very much more numerous the staminate flowers are than the fertile ones.

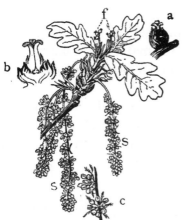

438. — Twig of oak with both kinds of flowers: *f*, fertile flowers; *s, s,* staminate; *a,* pistillate flower, enlarged; *b,* vertical section of pistillate flower, enlarged; *c,* portion of one of the sterile aments, enlarged, showing the clusters of stamens.

312. Advantages of the Unisexual Arrangement. — The absence of parts in a flower is not necessarily a mark of low organization, but may be the result of adaptation to its surroundings. It has been proved by experiment that flowers will generally produce more vigorous and healthy seed when impregnated with pollen from a different plant of the same species, and unisexual flowers promote this result by making it impossible for any blossom to receive pollen from itself.

313. Suppression or Abortion of Organs. — Sometimes this advantage is secured by the suppression of one or the other set of organs in different flowers. In the pistillate flowers of the persimmon the aborted stamens are quite conspicuous, though entirely sterile, producing not a grain of pollen. Rudimentary (undeveloped) organs of this kind are very common and are a frequent cause of irregularity and want of symmetry, as was seen in the stamens of the cress family (Sec. 295). Suppressed stamens are a common characteristic of the great bilabiate group (Sec. 308), large numbers of species having only two or four, but these are often accompanied, as in the pentstemon, chelone, and figwort, by sterile filaments in a more or less aborted condition that carry out the law of symmetry indicated in the five-lobed corolla (Sec. 308). The filament and style are often wanting, so that the anther or the stigma becomes sessile. While it is usual to speak of the stamens and pistil as essential organs, it is really only the ovary and the anther, or more strictly speaking, the ovules and pollen that are absolutely essential. The style is merely an appendage for placing the stigma where it will be brought more easily into contact with the pollen, and may be of any length, from a foot or more, as in the "silk" of the Indian corn, to a mere line, or entirely absent, as in the poppy and some of the yuccas.

439, 440. — Abortive stamens (*after* GRAY): 439, corolla of *Pentstemon grandiflorus* laid open, with its four stamens, and a sterile filament in the place of the fifth stamen; 440, corolla of catalpa laid open, with two perfect stamens and the vestiges of three abortive ones.

The study of these rudimentary or discarded organs helps to explain many deviations in the structure of flowers that would otherwise be very puzzling, and by their aid we can often reconstruct the plan of a flower that seems to have lost all conformity to the type.

314. Cleistogamic (*closed*) **Flowers** are so called because they never unfold, but are pollinated in the bud. Common examples are the inconspicuous closed flowers, on very short peduncles, concealed under the leaves of most violets. Sometimes, as in the fringed polygala, they are borne on underground stems and never rise above ground at all. The corolla is usually wanting and the stamens and pistil are greatly reduced, but they are much more prolific than ordinary blossoms.

315. Transformations.— Instead of suppression, organs frequently undergo an alteration into something else by

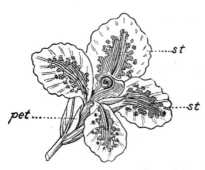

441. — Staminodia, transformed stamens of calla simulating petals: *pet*, petals; *st*, staminodia.

which their nature is greatly obscured. Conspicuous instances are the brilliant *staminodia*, or altered stamens of the canna, that simulate petals (Fig. 441), and the four large white bracts, usually mistaken for a corolla, that surround the flower clusters of the dogwood. In the cereus and other cactuses, bracts may be found in all stages of transition, from spines or scales to the most gorgeous of corollas. The rose, camellia, and water lily furnish other instances of the same kind; and in fact, examples of the transition of almost any organ into another may be observed by one who will take the trouble to look for them.

316. Appendages of the Corolla. — An appendage attached to the inner face of the corolla, like the funnel-shaped or bell-shaped projection within the perianth of daffodils and jonquils and others of the amaryllis family, to which they belong, is called a crown. It is no part of the peri-

442. — Flower of a cactus (*cereus greggii*), showing transition from scales to petals.

anth proper, and does not interfere in any way with the symmetry of the flower. The crown of the passion flower, to which so much of its beauty is due, is composed of a ring of abortive filaments, brilliantly colored, that surround the base of the style.   In the milkweed (*Asclepias*) the crown itself is appendaged with five little incurved horns.

317. Other Appendages. — Though appendages are most frequently connected with the calyx and corolla, they may attach to any part of the plant.   Figure 377 shows an appendaged anther; and the various appliances for dispersal furnish examples of appendaged fruits and seeds. When the appendage is so large as to inclose a whole seed, like the loose transparent sac around the seed of the water lily, and the brilliant scarlet pulp around the seeds of the strawberry bush (*Euonymous americanus*), it is called an *aril;* can you think of a use for it?

318. Use of Appendages. — The offices of these appendages are as varied as the appendages themselves.   They may be, as in the case of hairy filaments, to protect the pollen from crawling insects; to keep out rain, dew, or frost; to retain or to shed moisture; to secrete honey, as in the spurs and sacs of the violet and larkspur, or in other ways to attract and repel insects that aid or hinder the dispersal of pollen.   As they are generally the result of special adaptations on the part of the plant to its surroundings — more particularly with regard to insect pollination — they are usually indicative of an advanced stage of floral development.

PRACTICAL QUESTIONS

1. Why does a strawberry bed sometimes fail to fruit well, although it may flower abundantly?   (310, 311.)

2. Are berries found on all sassafras trees? on all buckthorns? hollies?

3. Would a solitary hop vine produce fruit? A solitary ash tree?

4. Why is a mistletoe bough with berries on it so much harder to find than one with foilage merely?   (310, 311.)

5. Explain the nature and use of the appendages in such of the plants named below as you can obtain; crown of the maypop, jonquil, milkweed; spurs of the columbine, tropæolum, jewel weed, etc.; bracts of the dogwood and poinsettia; spathe of Jack-in-the-pulpit and other arums.

NATURE AND OFFICE OF THE FLOWER

MATERIAL. — Any kind of large flower may be used ; those of the hollyhock, okra, cotton, hibiscus, or others of the mallow family are recommended, as their pollen grains are large enough to be observed fairly well with a hand lens. The cultivated Syrian hibiscus is the one used in the text.

319. Flower and Leaf. — We have seen that the venation of petals and sepals corresponds in a general way with that of foliage leaves of the class to which they belong, and that their arrangement around their axis is analogous to the arrangement of foliage leaves on the branch. We learned also, in our study of inflorescence, that flowers and flower buds occur only in the same positions where leaf buds occur, and that they are subject to the same laws of arrangement and growth.

320. Transformation of Organs. — In our study of fruits we saw that the carpels of the ovary are merely transformed leaves. We learned, also, in our study of leaves, something about the wonderful transformations that these organs are capable of undergoing ; and lastly, we have found some of these transformations taking place under our eyes in the leaflike sepals and petal-like filaments of the star-of-Bethlehem, in the bracts of the cactus, the scales of winter buds, and numerous other instances recorded in the preceding pages.

It must not be supposed, however, than an organ is ever developed as one thing and then deliberately changed into something else. When we speak loosely of one organ being transformed into another, the meaning is merely that it has developed into one thing instead of into something else that it was equally capable of developing into.

321. The Flower a Transformed Branch. — For the reasons mentioned, the flower is regarded by botanists as merely a branch with transformed leaves and the internodes indefinitely shortened so as to bring the successive cycles into close contact, the whole being greatly altered and specialized to serve a particular purpose.

322. The Course of Floral Evolution. — With this conception of the nature of the flower we can readily see that the less specialized its organs are and the more nearly they approach in structure and arrangement to the condition of an undifferentiated branch, the more primitive and undeveloped the type to which it belongs. On the other hand, if the parts are highly specialized and widely differentiated from the crude branch, a proportionately high stage of floral evolution is indicated.

323. Office of the Flower. — The one object of the flower is the production of fruit and seed, and all its wonderful specializations and variations of form and color tend either directly or indirectly to that end.

324. Fertilization. — It was stated in Section 286 that no seed can be developed unless some of the pollen reaches the stigma, but even this is not sufficient unless the process known as fertilization takes place. The exact nature of this process it is not easy to explain without going into details beyond the scope of this work, but a good general idea of fertilization may be obtained by referring to Figure 443 in connection with a study of the pollinated pistil of some large flower, like the hollyhock or hibiscus.

325. The Pollen Tubes. — Obtain if possible the flower of a Syrian hibiscus (okra will answer nearly as well) that has begun to close up, or to change color, and compare the stigma with that of a freshly opened flower. What difference do you observe in the pollen grains adherent to each? The yellow, withered look of the former is due to the fact that they have begun to germinate on the moist surface of the stigma; that is, to send down little tubes into its substance (Fig. 443, *i*), and the nourishment contained in the grain is being used up, just as the endosperm of the seed is used up when the embryo begins to germinate. (The germination of pollen, however, means something very different from the germination of the seed, and must not be confounded with it.) The pollen tube continues to elongate until it passes down through the base of the style into the ovary (Fig. 443, *m*).

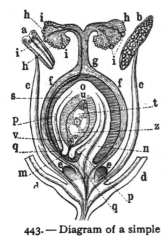

443.—Diagram of a simple flower, showing course of the pollen tube: *a*, transverse section of an anther before its dehiscence; *b*, an anther dehiscing longitudinally, with pollen; *c*, filament; *d*, base of floral leaves; *e*, nectaries; *f*, wall of carpels; *g*, style; *h*, stigma; *i*, germinating pollen grains; *m*, a pollen tube which has reached and entered the micropyle of the ovule; *n*, funicle of ovule; *o*, its base; *p*, outer integument; *q*, inner integument; *s*, nucellus of ovule; *t*, cavity of the embryo sac; *u*, its basal portion with antipodal cells; *v*, synergidæ; *z*, oösphere.

326. Course of the Pollen Tube. —The time taken for the tube to penetrate to the ovary varies in different flowers according to the distance traversed and the rate of growth. In the crocus it takes from one to three days, in the spotted calla (*Arum maculatum*), about five days, and in orchids, from ten to thirty days. In the hibiscus and many others of

444. — A pollen grain emitting a tube (magnified).

the mallow family, we know that it can not well exceed twenty-four hours, as the corolla usually falls away on the evening of the day on which it expanded, carrying the style and stamens with it, so that if the pollen tube had not reached the ovary by that time it could never get there at all. Sometimes the pistil is hollow, affording a free passage to the pollen tube; in other cases it is solid and the growing tube eats its way down, as it were, feeding upon the substance of the pistil as it grows. How is it in the flower you are examining? In some orchids the pollen tubes can be seen by the unaided eye, massed together within the thickened style, looking like a strand of fine white floss. It takes a grain of pollen to fertilize each ovule, and where more than one seed is produced to a carpel, as is commonly the case, at least as many pollen tubes must find their way to each cell of the ovary as there are ovules — provided all are fertilized.

327. Formation of the Seed. — When a pollen tube has penetrated to the ovary it next enters one of the ovules, usually through the micropyle (Fig. 443, *m*). There it penetrates the wall of a baglike inclosure called the embryo sac (Fig. 443, *u*, *t*, *z*), where a series of changes takes place too intricate to be described here, by which a fusion is brought about between a portion of the contents of certain cells emitted by the pollen tube and a large cell contained in the embryo sac, known as the germ cell, or egg cell (Fig. 443, *z*.). The fusion of these two bodies is what constitutes fertilization. The cell formed by their union finally develops into the embryo and the other contents of the sac into the endosperm, and the ripened ovules become the seeds.

328. Stability of the Process of Fertilization. — The processes of fertilization and reproduction are very obscure and difficult to understand without a degree of skill in the manipulation of the microscope and a knowledge of technical details that the ordinary observer can seldom acquire. The phenomena that characterize them, however, are the most uniform and stable of all the life processes, varying little not only in different species and orders, but throughout the whole vegetable kingdom. For this reason they furnish a more reliable standard for judging of the real affinities of plants than mere external resemblances, which are more liable to variation and may often be accidental, and so they have been chosen by botanists as the ultimate basis for the classification of plants.

329. Embryology. — The study of the developing ovule, known as embryology, is a comparatively recent branch of science, and has resulted in overturning many of the ideas of the older botanists and the abandonment of many of the established terms, which would now be misleading because they were founded upon false assumptions. This has led to a most unfortunate confusion in botanical terminology, the compensation for which lies in the hope that as investigation brings new truths to light greater clearness and certainty will grow out of the temporary disorder.

FIELD WORK

Look for examples of transition from one organ to another. These are particularly apt to occur in the so-called double flowers of the garden, and in those generally that have any of their organs indefinitely multiplied. Examine bracts and bud scales of different kinds, the carpellary leaves of leaflike follicles, such as those of the Japan varnish tree, milkweeds, columbine, and all sorts of vegetable monstrosities, which will generally be found to result from transformations of some sort. Study the numerical plan of some of the commonest flowers of your neighborhood; note the arrangement and consolidation of their organs, and determine their relative place in the evolutionary scale.

Make a list of all the outdoor plants, both wild and cultivated, that are found blooming in your neighborhood, keeping a record of the earliest specimens of each as you find them. The best way is to keep a sort of daily calendar, and at the end of each month give a summary of all the species found in bloom during that period. In this way a fairly complete annual record of the flowering time of the different plants for that vicinity will be obtained. The record should be kept up the whole year round. Don't stop in winter, but go straight on through the coldest as well as the hottest season, and you will make some surprising discoveries, especially if the record is kept up year after year. Give the common name of each plant, adding the botanical one if you know it. Any facts that you may know or may discover in regard to particular plants, such as their medicinal or other uses, their poisonous or edible properties, the insects that visit them, and in the case of weeds, their origin and introduction, will greatly enhance the interest and value of the record.

POLLINATION

MATERIAL. — This subject must be studied in the field and garden; no special directions for seeking material are needed.

330. Prevention of Self-Pollination. — The most interesting chapter in the history of plant life is that relating to the conveyance of pollen from the anther to the stigma.

It was recognized by the older botanists that this transfer was necessary to the production of fruit, but they were puzzled for nearly two hundred years by the fact that many flowers seem to be constructed as if on purpose to defeat this object. In our examination of the iris, for instance, it was seen that the anthers lie under the broad

445   446

445, 446.— Flower of fireweed (*Epilobium angustifolium*) (GRAY): 445, with mature stamens and immature pistil; 446, the same a few days older, with expanded pistil after the anthers have shed their pollen.

divisions of the style in such a manner that the pollen from them can not possibly reach the stigma without external agency; and in all monœcious and diœcious plants, self-pollination is clearly impossible. In other cases, of which the cone flower (*Rudbeckia*) and the common sage furnish examples, the anthers and stigma in the same flower do not mature together, thus producing the same effect as if they were unisexual.

331. Dimorphism is an expression for denoting a condition in which the stamens and pistils are of different relative lengths in different flowers of the same species, the stamens being long and the pistils short in some, the pistils long and the stamens short in others. Flowers of this sort are said to be dimorphous, or dimorphic, that is, of two forms; and some

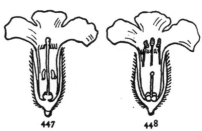

447   448

447, 448.— Flower of pulmonaria: 447, long styled; 448, short styled.

species are even *trimorphic*, having the two sets of organs long, short, and medium, respectively, in different indi-

viduals.   Examples of dimorphic flowers are the pretty
little bluets (*Houstonia cærulea*), the partridge berry (*Mitchella repens*), the swamp loosestrife (*Lythrum lineare*),

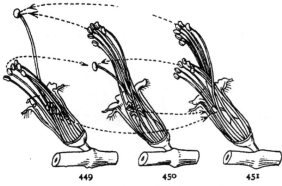

and the English
cowslip. Of trimorphic flowers
we have examples in the wood
sorrel, and the
spiked loosestrife
(*Lythrum salicaria*) of the gardens.   These
flowers were a
great puzzle to

449   450   451
449-451.—Three forms of *Lythrum salicaria*.

botanists until the celebrated naturalist, Charles Darwin,
proved by a series of careful experiments that the seed produced by pollinating a dimorphous flower with its own pollen,
or with pollen from a
flower of similar form, are
of very inferior quality to
those produced by impregnating a long-styled
flower with pollen from
a short-styled one, and
vice versa.

332. Wind Pollination.
—But the problem is
only half solved when a
plant has been rendered
incapable of impregnating itself.   Crosspollination, that is, the
transfer of pollen from a
separate flower or plant,

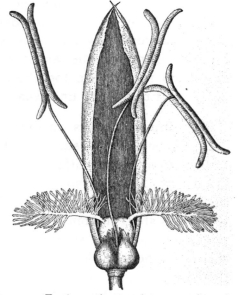

452.—Feathery stigmas of a grass adapted
to wind pollination.

has been rendered necessary, and provision must now be
made for the transportation.   In many cases, of which the

pine, Indian corn, oaks, ragweed, and grasses of all sorts afford abundant examples, this is accomplished by the wind. This is a very clumsy and wasteful method, however, for so much pollen is lost by the haphazard mode of distribution that the plant is forced to spend its energies in producing a vast amount more than is actually needed, and great masses of it are frequently seen in spring floating like patches of sulphur on ponds and streams in the neighborhood of pine thickets. Wind-pollinated flowers are called by botanists *anemophilous*, a word meaning "wind-loving." Like those that are self-pollinated, they are generally very inconspicuous, devoid of odor and of all attractions of form or color, because they have no need of these allurements to attract the visits of insects.

333. Insect Pollination. — A more economical method of securing pollination is through the agency of insects. In probing around for the nectar or the pollen upon which they feed, these busy little creatures get themselves dusted with the fertilizing powder, which they unconsciously convey from the stamen of one flower to the pistil of another. Insects usually confine themselves, as far as possible, to the same species during their day's work, and since less pollen is wasted in this way than would be done by the wind, it is clearly to the advantage of a plant to attract such visitors, even at the expense of a little honey, or of a liberal toll out of the pollen they distribute.

Flowers that have adapted themselves to insect pollination are said to be *entomophilous*, insect lovers, and all their various attractions of form, color, and odor have been developed, not for the gratification of man, as human arrogance and self-conceit have so long asserted, but as notifications to their insect guests that the banquet of nectar is spread.

334. Special Partnerships. — Some plants have adapted themselves to the visits of one particular kind of insect so completely that they would die out if that species were to become extinct. The well-known alliance between red

clover and the bumblebee was brought to light a **few**
years ago when the plant was first introduced into Aus-

tralia. It grew luxuriantly and blossomed
profusely, but would never set seed till the
bumblebee was introduced to keep it company.

The most remarkable of these partnerships,
perhaps, yet observed by naturalists, is that
which exists between the little *pronuba*, or
yucca moth, and the flowering yuccas, of
which the bear's grass and Spanish bay-

453.—Pod of
*yucca angusti-
folia* pierced by
the *Pronuba
yuccasella*.

onet of our
old fields and
roadsides are
familiar ex-

amples. If any of these
plants grow in your neighborhood, examine the pods
and observe that none of
them are perfect, but all
show a constriction at or

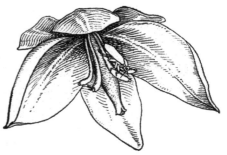

454.—Moth resting on yucca blossom.

near the middle, such as is sometimes seen in the sides
of wormy plums and pears. This is caused
by the larvæ of the moth, which feed upon
the unripe seeds. If you will look under
the nodding perianth of a yucca blossom
(Fig. 454), you will see that the short stamens are curved back from the pistil in
such a manner that under ordinary circumstances, not a grain of the pollen can fall
upon it except by the rarest accident. But
the yucca moth is a good farmer as well
as a provident mother, and as soon as she

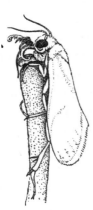

455.—Pronuba
pollinating pistil
of yucca.

has deposited her eggs in the seed vessel,
takes care to provide a crop of food for her
offspring by gathering a ball of pollen in
her antennæ and deliberately plastering it over the stigma
(Fig. 455). In this way she insures the perfecting of the

fruit and the proper nourishment of her children. When the eggs are hatched the larvæ feed upon the unripe seeds for a time, but it is rare that more than a dozen or two are destroyed in a pod, so that, after all, the plant pays only a moderate commission for the service rendered.

An equally interesting partnership exists between the Smyrna fig and the little insect, Blastophaga, an account of which may be found in the Year Book of the Department of Agriculture for 1900. In these cases the mutual dependence is so complete that neither the plant nor the animal could exist without the other.

335. Protective Adaptations. — Where plants have adapted themselves to insect pollination it is, of course, important to shut out intruders that would not make good carriers. In general, small, creeping things like ants and plant lice are not so efficient pollen bearers as winged insects, and hence the various devices, such as hairs, sticky glands, scales, and constrictions at the throat of the corolla, by means of which their access to the pollen is prohibited. To this class of adaptations belong the hairy filaments of the spiderwort, the sticky ring about the peduncles of the catchfly, the swollen lips of the snapdragon, the scales or hairs in the throat of the hound's-tongue, the velvet petals of the partridge berry, etc.

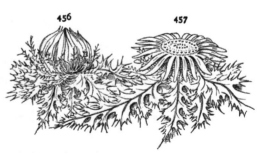

456, 457. — Protection of pollen in the thistle: 456, position at night or in wet weather; 457, position in sunshine.

Of flowers that are pollinated by night moths, some close during the day, as the four-o'clock and the evening primrose; and *vice versa*, the morning-glory, dandelion, and day flower (*Commelyna*), unfold their beauties only to the sun. For similar reasons, night-blooming flowers are generally white or very light colored, and shed their fra-

grance only after sunset.   A nodding position is assumed
by many flowers at night or during a shower to keep the
pollen from being injured by rain and dew.

458, 459. — A bell flower: 458, position in daylight; 459, position at night,
or during wet weather.

336. Fraud and Robbery. — The secretion of honey by
flowers is a very common means of attracting insect visit-

460. — Tubu-
lar blossom of
Acleisanthes
longiflora.

ors.   In general, plants that have very long,
tubular corollas, like the trumpet honeysuckle
(*Lonicera sempervirens*), and trumpet vine,
are reserving their sweets for humming birds
and long-tongued moths and butterflies.
Acleisanthes, a plant of the four-o'clock
family that grows along our Mexican border
(Fig. 460), has a tube from twelve to fourteen centimeters long (about five and one
half inches).   Yet even deeper corollas than
this can be explored by a humming bird of
South America, which has a bill that sometimes reaches the length of fifteen centimeters (about six inches), and a tongue that
can be protruded nearly as far again (Fig.
461).   It is not uncommon, however, to find

such corollas with a hole in the tube near the base, made by thieving bees and wasps which thus get at the honey surreptitiously, without paying their tribute of pollen. On the other hand, plants like the carrion flower, and skunk cabbage seem to practice

461.— Head and bill of sword bird (*Docimastes ensiferus*).

a kind of fraud upon flesh flies by imitating the colors and odors of the garbage upon which such creatures feed.

337. Experiments. — An instructive experiment may be made with regard to the color preferences of insects by putting a drop or two of syrup on bits of glass and laying them on paper of different colors in the neighborhood of a beehive or other place frequented by insects, and observing which color seems to attract them most. Similar experiments may be made with perfumes and flavorings.

338. Color, being a very variable and unstable quality, is of little use in classifying flowers, yet it is interesting to know that all their endless variations of hue are confined approximately within certain limits. Nobody has ever seen a blue rose or a yellow aster, and though the florist's art is constantly narrowing the application of this law, it still remains true that in a state of nature certain colors seem to be associated together in the floral art gamut. Yellow is considered by botanists the simplest and most primitive color in flowers, and blue the latest and most highly evolved. Yellow, white, and purple, in the order named, are the commonest flower colors in nature; blue the rarest.

PRACTICAL QUESTIONS

1. Why do the flowers of oak, willow, and other wind-fertilized plants generally appear before the leaves? (332.)

2. Can you account for the "showers of sulphur" frequently reported in the newspapers? (332.)

3. Do you see any connection between the feathery stigmas of most grasses and their mode of pollination? (332.)

4. Why are wind-fertilized plants generally trees or tall herbs?

5. If March winds should cease to blow, would vegetation be affected in any way?

6. Can you trace any connection between the winds and the corn crop?

7. Is it good husbandry to plant different varieties of corn, or other grain in the same field?

8. Why do the seeds of fruit trees so seldom produce offspring true to the stock? (333.)

9. Would you place a beehive near a field of buckwheat? Of clover? Near a strawberry bed? In a peach orchard? Near a fig tree? Under a grape arbor?

10. Why are very conspicuous flowers like the camellia, hollyhock, and pelargoniums so frequently without odor?

11. Why is the wallflower "sweetest by night"? (335.)

12. What advantage can flowers like the morning-glory gain by their early closing? (335.)

13. Of what use to the cotton plant, Japan honeysuckle, and hibiscus is the change of color their blossoms undergo a few hours after opening? (335.)

14. Why does the Japan honeysuckle, that has run wild so abundantly in many parts of our country, produce so few berries?

15. If the trumpet vine grows in your neighborhood, examine a number of corollas and account for the dead ants found in them. Try to account also for the large hole (sometimes three quarters of an inch in diameter) often found near the base of the tube. (336.)

16. Do you see any connection between the greater freshness and beauty of flowers early in the morning and the activity of bees, birds, and butterflies at that time?

17. The flowers most frequented by humming birds are the trumpet honeysuckle, cardinal flower, trumpet vine, horse mint (*Monarda*), wild columbine, canna, fuschia, etc.; what inference would you draw from this as to their color preferences?

FIELD WORK

The subject is itself so suggestive that it is hardly necessary to do more here than append a list of some of the plants which it would be interesting to examine with reference to their mode of pollination.

The orchids present the most wonderful adaptations for insect pollination, of all the vegetable kingdom, but they are rare and difficult to be obtained, so it is better to look for specimens nearer home. In neighborhoods where the pogonia, the purple and yellow fringed orchis, or the moccasin flower (*Cypripedium*) are found, they should, of course, receive attention. Some more easily obtainable specimens are:

Wallflower	Cheiranthus cheiri.
Bouncing Bet	Saponaria officinalis.
Columbine	Aquilegia vulgaris.
Monkshood	Aconitum napellus.
Larkspur	Delphinium (various species).
Barberry	Berberis vulgaris.
Mignonette	Reseda odorata.
Pansy	Viola tricolor.
Syrian Hibiscus . . .	H. syriacus.
Cotton	Gossypium (various kinds).
Nasturtium	Tropæolum majus.
Touch-me-not . . .	Impatiens (various species).
Wood sorrel	Oxalis (various species)
Horse-chestnut	Æsculus hippocastanum.
Buckeye	Æsculus pavia, flava, parviflora.
Pea	Pisum (various species).
Bean	Phaseolus (various species).
Ground nut	Apios tuberosa.
Vetch	Vicia.
Wistaria	Wistaria.
Black locust	Robinia pseudacacia.
Clover	Trifolium (various species).
Apple, pear	Pyrus.
Peach	Prunus persica.
Loosestrife	Lythrum salicaria.
Maypop	Passiflora incarnata.
Gourds, squashes, etc. . .	Cucurbitaceæ (various kinds).
Trumpet honeysuckle . .	Lonicera sempervirens.
Japan honeysuckle . . .	Lonicera Japonica.
Partridge berry	Mitchella repens.
Cone flower	Rudbeckia.
Dandelion	Taraxacum officinale.
Ox-eye daisy	Chrysanthemum leucanthemum.
Bell flower	Campanula rapunculoides.
Mountain laurel . . .	Kalmia latifolia.
Andromeda	A. ligustrina.
Primrose	Primula officinalis, P. grandiflora.
Persimmon	Diospyros virginiana.
Lilac	Syringa vulgaris.
Periwinkle	Vinca major, V. minor.
Milkweed	Asclepias (various species).
Snapdragon	Antirrhinum majus.
Lousewort	Pedicularis canadensis.
Trumpet vine	Tecoma radicans.
Horse balm	Collinsonia canadensis.

Dead nettle	.	Lamium amplexicaule, L. album.
Sage	.	Salvia officinalis and other species.
Catmint	.	Nepeta cataria.
Iris	.	Iris (various kinds).
Carrion flower . . .	.	Smilax herbacea.
Bear's grass . . .	.	Yucca filamentosa.
Spanish bayonet . .	.	Yucca aloifolia.
Lily of the valley . .	.	Convallaria majalis.
Day lily	.	Hemerocallis fulva.

PRACTICAL EXPERIMENTS

Experiments should be made by enveloping buds of various kinds in gauze, so as to exclude the visits of insects, and noting the effect upon the production of fruit and seed. Envelop a cluster of milkweed blossoms in this way and notice how much longer the flowers so protected continue in bloom than the others; why is this? Try the same experiment upon the blooms of cotton and hibiscus if you live where they grow, and see whether the characteristic change in color occurs in flowers from which insects have been excluded and whether good seed pods are produced by them. Try the effect upon fruit production of excluding insects from clusters of apple, pear, and peach blossoms.

ECOLOGICAL FACTORS

339. Definition. — By ecology is meant the relations of plants to their surroundings.   These may be classed under three general heads : their relations to inanimate nature, to other plants, and to animals.   The subject has been touched upon repeatedly in the foregoing pages, and, in fact, it is impossible to treat of any branch of botany without some reference to it.   All that was said about the adjustment of leaves for light and moisture, and their adaptations for protection and food storage, the devices for fruit and seed dispersal, etc., really belong to ecology, while Sections 330–338, about pollination, may be regarded as a very imperfect review of the ecology of the flower in relation to the insect world.

340. Symbiosis. — Associations for mutual help, like those described in Sections 330–338, between certain plants and their insect visitors, have been included by botanists under the general term, *symbiosis*, a word which means " living together."   In its broadest sense symbiosis refers to any sort of dependence or intimate organic relation between different kinds of individuals, and so may include the climbing and parasitic habits ; but it is more properly restricted to cases where the relation is one of mutual benefit.   It may exist either between plants of one kind with another, between animals with animals, or between plants and animals, as in the case of the clover and bumblebee, and the yucca and pronuba.

The occurrence of the root tubercles on certain of the leguminosæ (Sec. 198) is a clear case of symbiosis, the microscopic organisms in the tubercles getting their food from the plant and at the same time enabling it to get food for itself from the air in a way that it could not otherwise do.

341. Relations with Inanimate Nature. — But it is to the relations of plants with inanimate nature, and their grouping into societies under the influence of such conditions, that the term "ecology" is more strictly applied. The external conditions that lead to the grouping are called *ecological factors*. The most important of these are temperature, moisture, soil, light, and air, including the direction and character of the prevailing winds. Each of these factors is complicated with the others and with conditions of its own in a way that often makes it difficult to determine just what effect any one of them may have in the formation of a given plant society.

342. Temperature, for instance, may be even and steady like that of most oceanic regions, or it may be subject to sudden caprices and variations like the "heated terms" and "cold snaps" that afflict our northern and southern States respectively every few years. We must remember, too, that it is not the average temperature of a climate but its extremes, especially of cold, that limit the character of vegetation.

Temperature probably has more influence than any other factor in the distribution of plants over the globe, but it can have little or no effect in evolving local differences in vegetation because the temperature of any given locality, except on the sides of high mountains, will ordinarily be practically the same within a circuit of many miles.

343. Moisture, again, may be of all degrees, from the superabundance of lakes and rivers and standing swamps, to the arid dryness of the desert, and the water may be still and sluggish, or in rapid motion. It may exist more or less permanently in the atmosphere, as in moist climates like those of England and Ireland, where vegetation is characterized by great verdure, or it may come irregularly in the form of sudden floods, or at fixed intervals, causing an alternation of wet and dry seasons. Moreover, the moisture of the soil or the atmosphere may be impregnated with minerals or gases which may affect the vegetation independently of the actual amount of water absorbed.

344. Light may be of all degrees of intensity, from the blazing sun of the treeless plain to the darkness of caves and cellars where nothing but mold and slime can exist. Between these extremes are numberless intermediate stages; the dark ravines on the northern side of mountains; the dense shade of beech and hemlock forests; the light, lacy shadows of the pines; each characterized by its peculiar form of vegetation. Absence of light, too, is usually accompanied by a lowering of temperature and reduction of transpiration, factors which tend to accentuate the difference between sun plants and shade plants, giving to the latter some of the characteristics of aquatic vegetation. Generally, the tissues of these are thin and delicate, and having no need to guard against excessive transpiration they wither rapidly when broken.

345. Winds affect vegetation not only as to the manner of seed distribution, as in the case of tumbleweeds and winged fruits, but directly by increasing transpiration, and necessitating the development of strong holdfasts in plants growing upon mountain sides and in other exposed situations.

462. — A red cedar grown under natural conditions.

463. — A red cedar grown in a barren, wind-beaten situation.

The nature of the region from which they blow — whether moist, dry, hot, cold, etc., is also an important factor. In a district open to sea breezes, live oaks, which require a salt atmosphere, may sometimes be found as far as a hundred miles from the coast.

346. Soil is perhaps the most interesting of these factors to the farmer, because it is the one that he has it most largely in his power to modify. It is to be viewed under two aspects : first, as to its mechanical properties, whether soft, hard, compact, porous, light, heavy, etc.; secondly, as to its chemical composition and the amount of plant food contained in it. The first can be regulated by tillage and drainage, the second by a proper use of fertilizers.

Under mechanical structure is included also the power of absorbing and retaining water. A good absorbent soil, *i.e.* sand, or gravel, is not apt to be a good retainer, while clay and marl, that absorb slowly, retain well.

347. Experiment. — Take a few handfuls of each of the different kinds of soil in your neighborhood, free them as thoroughly as possible from all traces of vegetation, place separately in small earthen pots or saucers and keep them well moistened. Pull up the seedling plants that appear in each, and keep a list of them as long as any continue to come up. What inference would you draw from the number produced in each pot as to the productiveness of the different soils? Could all the seedlings have lived if they had been left to grow where they came up? What becomes of the majority of seedlings that germinate in a state of nature?

PRACTICAL QUESTIONS

1. Is the relation between man and the plants cultivated by him a symbiosis?

2. Why is it that plants of the same, or closely related species, are found in such different localities as the shores of Lake Superior, the top of Mt. Washington, and the Black Mountains in North Carolina? (342.)

3. Which of the five ecological factors described in Sections 341–346 has probably influenced their distribution? (342.)

4. What is the prevailing character of the soil in your neighborhood?

5. Is your climate moist or dry? Warm or cold?

6. Can you trace any connection between these factors and the prevailing types of vegetation?

PLANT SOCIETIES

MATERIAL. — A specimen of pipewort (*Eriocaulon*), Sagittaria, pondweed, or other succulent water plant, and a cactus of some kind. The common prickly pear (*Opuntia*) is the one used in the text. City schools should have a small aquarium; a few water plants can be kept in jars.

348. Principles of Subdivision. — Plants group themselves into societies not according to their botanical relationships, but with regard to the predominance of one or more of the ecological factors that influence their growth. Sometimes one or two species will take practical possession of large areas, like the coarse grasses that spread over certain salt marshes, or the pines that formerly constituted the sole forest growth over extensive regions in North Carolina and Maine. But more usually we shall find a great diversity of forms brought together by their common requirements as to shade, soil, moisture, etc. These societies are, of course, purely artificial, and any of the factors named in Sections 341–346, or others of a different kind, may be made the basis of their classification. They might be grouped, for instance, according to the soil in which they grow, or according to origin, whether cultivated, wild, native, introduced, etc., as best suited the purpose of the classification in each case. The moisture factor, however, has been generally agreed upon by botanists as the one most convenient for ordinary purposes. Upon this principle plants are divided into three great groups : —

Hydrophytes, or water plants, those that require abundant moisture.

Xerophytes, or drought plants, those that have adapted themselves to desert or arid conditions.

Mesophytes, plants that live in conditions intermediate between excessive drought and excessive moisture. To this class belong most of our ordinary cultivated plants and the greater part of the vegetation of the globe.

Halophytes, "salt plants," is a term used to designate a fourth class, based not directly upon the water factor, but upon the presence of a particular mineral in the water or the soil, which they can tolerate. They seem to bear a sort of double relation to hydrophytes on the one hand and to xerophytes on the other.

349. Hydrophyte Societies. — These embrace a number of forms, from those inhabiting swamps and wet moors to the submerged vegetation of lakes and rivers. An exami-

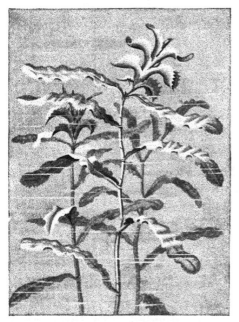

464. — A hydrophyte society of floating pond-weed.

465. — A water plant (*Sagittaria natans*), showing the slender, ribbonlike, submerged leaves, the broad, rounded, floating ones, and the very slightly developed root system.

nation of almost any kind of water plant will show some of the physiological effects of unlimited moisture. Take a piece of pondweed, or other immersed plant out of the water and notice how completely it collapses. This is because, being buoyed up by the water, it has no need to spend its energies in developing woody tissue. Floating and swimming plants will generally be found to have no root system, or only very small ones,

because they absorb their nourishment through all parts
of the epidermis directly from the medium in which they
live. That they may absorb
readily, the tissues are apt to
be soft and succulent and the
walls of the cells composing
them very thin. In some of
the pipeworts (*Eriocaulons*), the
cells are so large as to be
easily seen with the unaided
eye. If you can obtain one
of these, examine it with a lens
and notice how very thin the
walls are. Water plants also
contain numerous air cavities,

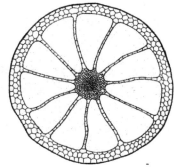

466. — Transverse section through
the stem of a hydrophyte plant
(*Elatine alsinastrum*), showing the
very large air cavities (GOODALE,
after REINKE).

and often develop bladders and floats, as in the common
bladderwort, and many seaweeds (Fig. 467).

Swamp plants, drawing their nourishment from the
loose soil in which they are anchored, and lacking the
support of a liquid medium, develop
roots and vascular stems. The roots

467. — Seaweed (*sar-
gassum*) with bladderlike
floats.

468. — A cypress trunk, showing enlarged base for
aëration.

of plants growing in swamps have difficulty in obtaining
proper aëration on account of the water, which shuts off
the air from them, hence they are furnished with large

air cavities, and the bases of the stems are often greatly enlarged, as in the Ogeechee lime (*Nyssa capitata*) and cypress, to give room for the formation of air passages. The peculiar hollow projections known as "cypress knees" are arrangements for aërating the roots of these trees.

350. Xerophyte Societies are adapted to conditions the reverse of those affected by hydrophytes. The extreme of these conditions is presented by regions of perennial drought like our western arid plains and the great deserts of the interior of Asia and Africa. Under these conditions

469. — "Switch plants" of the alkali desert, condensed into mere green skeletons of vegetation, and thus adapted to extreme xerophyte conditions.

plants have two problems to solve; to collect all the moisture they can and to keep it as long as they can. Hence, plants of such regions diminish their evaporating surface by reducing or getting rid of their foliage and compacting all their tissues into the stem, like the cactus (Sec. 209), or they compress their leaves into thick and fleshy forms fitted to resist evaporation and retain large amounts of moisture, as in the case of the yucca and century plant. They also frequently develop a thick, hard epidermis, or cover themselves with protective hairs and scales.

351. Examination of a Xerophytic Plant. — Examine a joint of the common prickly pear (*Opuntia*), if it grows in your neighborhood, or use a potted cactus, and give your reasons for regarding it as a stem and not as a leaf.

Notice how the spines are arranged on the surface, and if there are any fruits, buds, or flowers, where they occur. Peel off a little of the epidermis and observe its thick, horny texture. Cut a cross section through a joint about midway from base to apex and examine with a lens. Notice the thick layer of green tissue next the epidermis, and within that, a band of tough, woody fibers inclosing the soft pulpy mass that makes up the interior. (If the woody layer is not easily made out, allow your specimen to dry for about twenty-four hours, and it will become quite distinct.) Make a longitudinal section through the center of a joint and trace the course of the woody fibers; do they get any more abundant

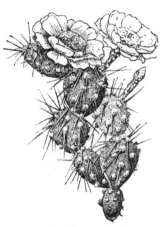

470. — A plant of opuntia, showing young branches and flowers from the nodes.

toward the base? Do any of them pass into the spine clusters? What do the spines represent? What is the use of the green layer just under the epidermis? Why is it so much more abundant in the cactus than in ordinary stems? Lay aside a section of a cactus plant, or a leaf of yucca, agave, or other fleshy xerophyte to dry and see how long it takes to lose its moisture. What would you conclude from this as to its retentive power?

352. Other Xerophyte Adaptations. — Plants exposed to periodic and occasional droughts frequently provide against hard times by laying up stores of nourishment in bulbs and rootstocks and retiring underground until the stress is over. This is known to botanists as the *geophilous*, or earth-loving habit. Others, as some of the lichens, and the little resurrection fern (*Polypodium incanum*), so common on the trunks of oaks and elms, make no resistance, but wither away completely during dry weather, only to waken again to vigorous life with the first shower.

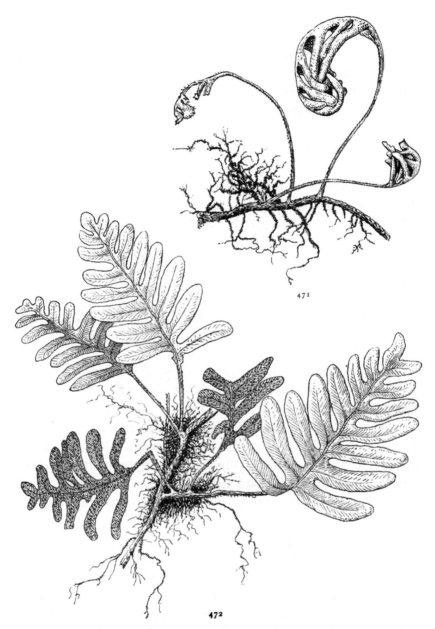

471, 472. — A resurrection fern: 471, in dry weather; 472, after a shower.

353. Mesophytes. — These embrace the great body of plants growing under ordinary conditions, which may vary from the liberal moisture of low meadows and shady forests to the almost xerophytic barrenness of dusty lanes and gullied hillsides.   The forms and conditions they present are so diversified that it will be impossible even to touch upon them all in a work like this, but they may be summed up under the two principal heads of open ground and woodland growth.   Under the first are included all cultivated grounds; fields, lawns, meadows, pastures, and roadsides, with their characteristic weeds, flowers, and grasses.   Under the second, all woods and copses with the shrubs and herbs that form their undergrowth.

354. Halophytes include plants growing by the seashore and the vegetation around salt springs and lakes and that of alkali deserts.   Seaweeds are in a sense halophytes, since they live in salt water, but as they are true aquatic plants and exhibit many of the peculiarities of hydrophytes in their mechanical structure, they are classed with them. The name halophyte applies more particularly to land plants that have adapted themselves to the presence of certain minerals, popularly known as salts, in the soil or in the atmospheric vapor.   If you have ever spent any time at the seashore, you can not fail to have been struck with the thick and fleshy habit exhibited by many of the plants growing there, such as the samphire, sea purslane (*Sesuvium*), and sea rocket (*Cakile*).   A form of goldenrod found by the seashore has thick, fleshy leaves, and is as hard to dry as some of the fleshy xerophytes.

Another characteristic of desert plants that is common also to seaside vegetation, is the frequent occurrence of a thick, hard epidermis, as in the sea lavender and saw grass.   The live oaks, trees that love the salt air and never flourish well beyond reach of the sea breezes, have small, thick, hard leaves, very like those of the stunted oaks that grow on the dry hills of California.   The presence of spines and hairs, it will be observed, is also very common; *e.g.* the salsola, the sea ox-eye, and the low primrose (*Œnothera humifusa*).   In other cases the leaf blades are so strongly involute or revolute (Sec. 60) as to make them appear cylindrical — an arrangement for protecting the stomata (Fig. 98) and preventing transpiration. All these, it will be observed, are xerophyte characteristics, and the object in both cases is the same — economy of moisture. The reason why such adaptations are necessary in halophyte plants is because the mixture of salt in the water of the soil increases its density so that it is difficult for the plant to absorb what it needs (Sec. 227). Hence, halophytes are in the condition of Coleridge's "Ancient Mariner"; with "water, water everywhere," they are practically living under xerophyte conditions.

PRACTICAL QUESTIONS

1. Why do florists always cultivate cactus plants in poor soil? (350.)

2. What would be the effect of copious watering and fertilizing on such a plant? (350.)

3. Why must an asparagus bed be sprinkled occasionally with salt? (348, 354.)

4. If a gardener wished to develop or increase a fleshy habit in a plant, to what conditions of soil and moisture would he subject it? (350, 354.)

5. What difference do you notice between blackberries and dewberries grown by the water and on a dry hillside?

6. Is there a corresponding difference between the root, stem, or leaves of plants growing in the two situations, and if so account for it?

7. When a tract of dry land is permanently overflowed by the building of a dam or levee, why does all the original vegetation die, or take on a very sickly appearance? (349.)

8. Should plants with densely hairy leaves be given much water, as a general thing? (354.)

9. A farmer planted a grove of pecan trees on a high, dry hilltop; had he paid much attention to ecology?

10. Give a reason for your answer.

FIELD WORK

Ecology offers the most attractive subject for field work of all the departments of botany. It can be studied anywhere that a blade of vegetation is to be found. In riding along the railroad there is an endless fascination in watching the different plant societies succeed one another and noting the variations that they undergo with every change of soil or climate.

Students in cities can study ecology in parks and public squares, in the vegetation that springs up on vacant lots, around doorsteps and area railings, and even between the paving stones of the more retired streets. A botanist found on a vacant lot near the public library in Boston over thirty different kinds of weeds and herbs, and in the heart of Washington, D.C., on a vacant space of about twelve by twenty feet, nineteen different species were counted. Even in great cities like London and New York, one occasionally recognizes among the rare weeds struggling for existence with the paving stones in out-of-the-way corners, some old acquaintance of fields and roadsides far away. Just where all these things came from, and how they got there, and why they stay there, will be interesting questions for city students to solve.

But the country always has been and always will be the happy hunting ground of the botanist. All the factors considered in the two preceding sections can hardly be found in any one locality, but mesophyte and hydrophyte conditions exist almost everywhere, and approximations to the xerophyte state can generally be found at some season in open, sandy, or rocky places, along the borders of dry, dusty roads, and on the sun-baked soil of old red hills and gullies.

If there are any bodies of water in your neighborhood (in cities, visit the artificial lakes in parks), examine their vegetation and see of what it consists. Notice the difference in the shape and size of floating and immersed leaves and account for it. Note the general absence of free-swimming plants in running water, and account for it. Note the difference between the swamp and border plants and those growing in the water, and what trees or shrubs grow in or near it. Compare the vegetation of different bogs and pools in your neighborhood, and account for any differences you may observe; why, for instance, does one contain mainly rushes, sedges, and cat-tails, another ferns and mosses, another sagittaria, boneset, water plantain, etc., and still another a mixture of all kinds? Compare the water plants with those growing in the dryest and barrenest places in your vicinity, note their differences of structure, and try to find out what special adaptations have taken place in each case.

Draw a map of some locality in your neighborhood that presents the greatest variety of conditions, representing the different ecological regions by different colored inks or crayons, or by different degrees of shading with the pencil.

X. SEEDLESS PLANTS

355. Order of Development. — All the forms that have hitherto claimed our attention belong to the great division known as Spermatophytes, or seed-bearing plants, sometimes designated also as Phanerogams, or flowering plants. They comprise the higher forms of vegetable life, and because they are more striking and better known than the other groups, they have been taken up first, since it is easier for ordinary observers to work their way backwards from the familiar to the less known.

• But it must be understood that this is not the order of nature. The geological record shows that the simplest forms of life were the first to appear and from these all the higher forms were gradually evolved. There is no sharp line of division between any of the orders and groups of plants, but the line of development can be traced through a succession of almost imperceptible changes from the lowest forms to the highest, and it is only by a study of the former that botanists have come to understand the true nature and structure of the latter.

It would be impossible, in a work like this, to attempt even a superficial view of the various divisions of seedless plants. Many of them are of microscopic size, and can not be studied without expensive laboratory appliances and skill in the manipulation of the microscope, which not everybody can possess. A short study of only a few typical forms will be attempted here, in order to make clearer some of the processes of plant life that have already been touched upon.

356. Classification. — Beginning with the lowest forms, seedless plants are grouped into three great orders, or classes.

357. I. Thallophytes, or thallus plants. This group takes its name from the thallus structure that characterizes its vegetation. What a thallus is will be better understood after a specimen has been examined. It may be stated, however, that the term is applied in general to the simplest kinds of vegetable structure, in which there is no differentiation of tissues, and no true distinction of root, stem, and leaves. While it is not peculiar to the thallophytes, it has

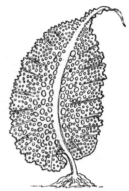

473. — A seaweed with broad expanded thallus.

attained its most typical development among them, and the name is therefore retained as distinctive of that group. It embraces two great divisions, the Algæ and Fungi. The first includes seaweeds and the common fresh-water-brook silks, pond scums, etc., besides numerous microscopic forms whose presence escapes the eye altogether, or is made known only by the discolorations and other changes they effect in the water.

To the fungi belong the mushrooms and puff balls, the molds, rusts, mildews, etc., and the vast tribe of microscopic organisms called bacteria, that are so active in the production of fermentation, putrefaction, and disease.

474. — Anthoceros, a liverwort with flat, spreading thallus.

358. II. Bryophytes, or moss plants. This group likewise contains two divisions, mosses and liverworts. Familiar examples of the latter are the marchantia, or umbrella liverwort (Figs. 500, 502), commonly found on the ground in cool bogs, and the flat, spreading plants, bearing somewhat the aspect of lichens,

475. — Scapania, a liverwort with leafy thallus, approaching the form of mosses and lycopodiums (from COULTER'S "Plant Structures").

except for their color, met with everywhere on wet rocks and banks around shady water courses.

Mosses are one of the best defined of botanical orders, and are too well known to need further specification here.

Bryophytes form a connecting link, or rather a chain of connecting links between the next group, pteridophytes, and thallophytes. The liverworts represent the more primitive division of the group, and in some of their forms approach so near the thallophytes that it does not take a botanist to recognize the relationship.

477. — A common fern (*Polypodium vulgare*).

359. III. Pteridophytes, or fern plants, include the three divisions of ferns, horsetails, and club mosses. They differ greatly in structure, but all possess a vascular system, a well-organized

476. — A common moss plant, with parts apparently divided into root, stem, and leaves, but with no true differentiation of tissues (from COULTER'S "Plant Structures").

system of root, stem, and leaves, and rank next to the spermatophytes in the order of development. They are frequently distinguished as the vascular cryptogams to differentiate them from the other two groups, cryptogams being a term sometimes used to designate the three orders of seedless plants. The distinction

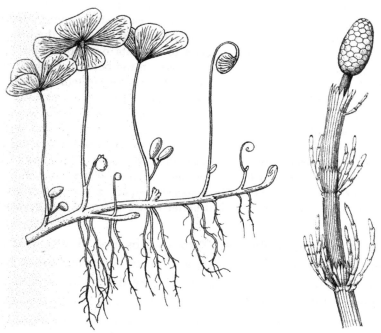

478. — A water pteridophyte, *Marsilia* (*after* GRAY).

479. — Part of the fruiting stem of a scouring rush, *Equisetum limosum* (*after* GRAY).

between vascular and non-vascular plants is relatively as important a one as that between vertebrates and invertebrates in the animal kingdom.

Just what these three great groups are, and what relation they bear to one another, will be better understood by the study of a typical specimen of each.

FERN PLANTS

MATERIAL. — Any kind of fern in the fruiting stage. The pretty little ebony fern (*Asplenium ebeneum*), and the Christmas fern (*Aspidium acrostichoides*) are common almost everywhere, the former on shady hillsides near the foot of rocks and stumps, or in the shadow of walls and fences ; the latter in rocky woods and along water courses

almost everywhere. City schools can supply themselves with specimens by cultivating a few ornamental ferns in the schoolroom. While
gathering specimens look along the ground under the fronds, or in
greenhouses where ferns are cultivated, among the pots and on the floor,
for a small, heart-shaped body like that represented in Figures 493, 494,
called a *prothallium*. It is found only in very wet places and care must
be taken in collecting specimens, as in their early stages the prothalli
bear a strong resemblance to certain liverworts found in the same
places. The best way is for each class to raise its own specimens by
scattering the spores of a fern in a glass jar, on the bottom of which is
a bed of moist sand or blotting paper. Cover the jar loosely with a
sheet of glass and keep it moist and warm, and not in too bright a light.
Spores of the sensitive ferns (*Onoclea*) will germinate in from two to
ten days, according to the temperature. Those of the royal fern
(*Osmunda*) germinate promptly if sown as soon as ripe, but if kept
even for a few weeks are apt to lose their vitality. The spores of
sensitive fern can be kept for six months or longer, while those of the
bracken (*Pteris*) and various other species require a rest before germinating, so that in these cases it it better to use spores of the previous
season.

360. Study of a Typical Fern. — Observe the size and
general outline of the fronds, and note whether those of
the same plant are all alike, or if they differ in any way,
and how. Observe the shape and texture of the divisions
or pinnæ composing the frond, their mode of attachment
to the rhachis, and whether they are simple, or notched or
branched in any way. Make a sketch, labeling the primary branches of the frond, *pinnæ* (sing. *pinna*), the
secondary ones, if any, *pinnules*, and the common stalk
that supports them, *stipe*. Note the color, texture, and
surface of the stipe. If any appendages are present, such
as hairs, chaff, or scales, notice whether they are most
abundant toward the apex or the foot of the stipe, or
equally distributed over its whole length. Cut a cross
section near the foot and look through your lens for the
roundish or oblong dots that show where the fibrovascular
bundles were cut through (Fig. 482). How many of them
do you see? Make a sketch and compare with your sectional drawings of the stems of monocotyledons and
dicotyledons; what differences do you notice? Which
does it resemble most?

Examine the mode of attachment of the stipes to their underground axis. Break one away and examine the scar.

Compare with your drawings of leaf scars and with Figure 274. Do the stipes grow from a root or a rhizoma? How do you know? Do you find any remains of leafstalks of previous years? How does the rootstock increase in length? Measure some of the internodes; how much did it increase each year? Cut a cross section and look for the ends of the fibrovascular bundles. Trace their course through several internodes. Do they run straight or they run straight or do they turn or bend in any way at the nodes? If so, where do they go?

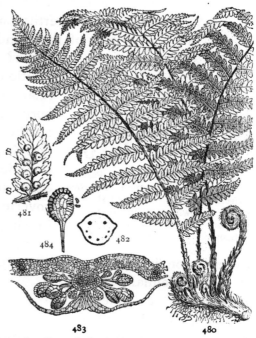

480–484. — A fern plant: 480, fronds and rootstock; 481, fertile pinna: *s, s,* sori; 482, cross section of a stipe, showing ends of the fibrovascular bundles; 483, a cluster of sporangia, magnified; 484, a single sporangium still more magnified, shedding its spores.

361. Veining. — Hold a pinna up to the light and examine the veining. Is it like any of the kinds described in Sections 37–40? This forked venation is a very general characteristic of the ferns. When the forks do not reticulate or intercross in any way, the veins are said to be free; are they free in your specimen, or reticulated?

362. Fructification. — Examine the back of the frond; what do you find there? (Most ferns bear many sterile fronds; care must be taken to secure some fruiting ones.) These dots are the *sori* (sing. *sorus*), or fruit clusters, and the fronds or pinnæ bearing them are said to be *fertile*. Are there any differences of size, shape, etc., between the fertile and sterile fronds of your specimen ? Between the fertile and sterile pinnæ ? On what part of the frond are the fertile pinnæ borne ? Notice the shape and position of the sori, and their relation to the veins, whether borne at the tips, in the forks, on the upper side (toward the margin) or the lower (toward the midrib). Look for a deli-cate membrane (*indusium*) covering the sori, and observe its shape and mode of attach-ment. (If the specimen under examination is a polypodium there will be no indusium ;

485. — Part of a fertile pinna of *polypodium* en-larged, showing the sori without indusium.

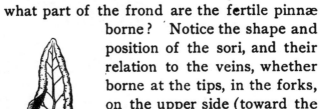

486. — Part of a pinna of *pellea* enlarged, showing indusium formed by the revolute margin.

if a maidenhair (*Adi-antum*), or a bracken (*Pteris*), it will be formed of the revolute margin of the pinna.) In lady fern (*Asplenium Filix-fœmina*), and Christmas fern (*Aspidium*), the sori frequently become con-fluent, that is, so close together as to appear like a solid mass. Sketch a

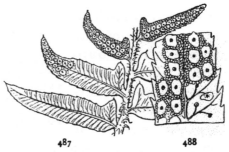

487

488

487, 488. — Christmas fern (*Aspidium*) : 487, part of a fertile frond, natural size ; 488, a pinna enlarged, showing the sori confluent under the peltate indusia.

fertile pinna as it appears under the lens, bringing out all the points noted.

363. The Spore Cases. — Look under the indusium at the cluster of little stalked circular appendages (Fig. 483). These are the *sporangia*, or spore cases, in which the reproductive bodies are borne. Seen under the micro-scope each sporangium looks like a little stalked bladder surrounded by a jointed ring (Fig. 484). At maturity the ring straightens itself out, ruptures the wall of the sporangium, and the spores are discharged with considerable

489   490   491   492

489-492. — Spores of pteridophytes, magnified: 489, a fern spore; 490, 491, two views of a spore of a club moss; 492, spore of a common horsetail (*Equisetum arveuse*).

force.   Compare the spores depicted in Figures 489–492, with the pollen grains in Figures 378–381.   Do you notice any resemblance?

364. Reproduction. — The spores are the reproductive bodies of ferns, and correspond in this respect to the seeds of spermatophytes, but their mode of reproduction is very different, or rather seems so, because here the process known as *alternation of generations* first becomes apparent to the eye, as we proceed from the higher plants to the lower.   The same thing occurs among seed plants also, but as it is there partly concealed within the seed, botanists first became acquainted with it through the study of spore-bearing plants, where it is more clearly revealed.   What is meant by it will be better understood after the life history of the ferns has been studied.

365. The Sporophyte. — The spores found in such abundance on the fertile pinnæ are all alike, and each one is capable of germinating and continuing the work of reproduction without the necessity of any such union as we saw taking place between the pollen and the ovule in the spermatophytes.   The plant or part of a plant that bears these reproductive bodies is called a *sporophyte*, or spore plant, and with its crop of spores makes up one generation.

366. The Prothallium. — When one of these spores germinates, it produces, not a fern plant like the one that bore it, but a small, heart-shaped body like that shown in

Figure 493, called a prothallium. Examine one of these bodies carefully with a lens. Observe that there are no veins nor fibrovascular bundles, and the whole body of the plant seems to consist of one uniform tissue. Some little rootlike hairs, called *rhizoids*, will be found growing on the under side, but these are shown by the microscope to be mere appendages of the epidermis in the nature of hairs, and not true roots. Such a body as this, in which there is no differentiation of parts, is what constitutes a thallus. It occurs in all kinds of plants under varying forms, and different names are given to it. In the ferns it is called a *prothallium*. In them it is generally short-

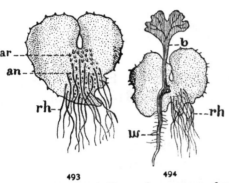

493, 494. — Prothallium of a common fern (*Aspidium*): 493, under surface, showing rhizoids, *rh*, antheridia, *an*, and archegonia, *ar*; 494, under surface of an older gametophyte, showing rhizoids, *rh*, and young sporophyte, with root, *w*, and leaf, *b* (from COULTER'S "Plant Structures").

lived and is important only in connection with the work of reproduction. Note its heart-shaped outline, and look just below the deep notch at the apex for certain little bottle-shaped bodies called *archegonia*. (They will probably appear under the lens as mere dots, or may not be visible at all.) These correspond to the pistils of seed plants. Lower down, among the rhizoids, or near the margin of the prothallium, are certain organs, called *antheridia*, corresponding to the stamens of spermatophytes.

367. The Gametophyte. — The reproductive cells contained in the antheridia and archegonia are called *gametes* and from them the prothallium is called a *gametophyte*, or gamete plant, in contradistinction to the sporophyte or spore plant. The gametes differ from ordinary spores in not being able to perform the work of reproduction directly by germination, but a pair of them must first unite and form another kind of spore called an *oöspore*, which is capable of germinating. It reproduces, however, not the simple thalluslike gametophyte from which it sprang, but the beautiful fern plant, or sporophyte, with its vascular system and complete outfit of vegetative organs — root, stem, and leaves.

368. Alternation of Generations. — We all know the meaning of the word generation as applied to the direct descendants of one organism from another, whether animal or plant. When two successive generations produce respectively ordinary spores and oöspores, and these different kinds of spores give rise to organisms unlike in structure or habits of life, there is said to be an alternation of generations. The generation which bears the simple spores (sporophyte) is said to be asexual; the one which produces the gametes and oöspores is sexual; that is, it requires the union of two separate bodies to produce a fertilized germ, or oöspore. Each generation, therefore, it will be observed, gives rise to its opposite, the asexual sporophyte producing the sexual gametophyte, or prothallium, and this in turn, through its gametes and oöspores reproducing the asexual sporophyte. The alternation in ferns may, in general, be expressed to the eye by a series of diagrams like those given below. The words in each line are synonyms of those immediately above or below them in the other lines, except it must be observed that, strictly speaking, it is not the antheridia and archegonia, but the spores or gametes contained in them that by their union produce the oöspore.

ern plant → Spores → Prothallium → ⟨ Archegonia / Antheridia ⟩ → Oöspore → Fern plan

porophyte → Spores → Gametophyte → ⟨ Gamete / Gamete ⟩ → Oöspore → Sporophyte

sexual gen. → Spores → Sexual gen. → ⟨ Gamete / Gamete ⟩ → Oöspores → Asexual ge

369. Advantages of Alternation. — This roundabout mode of reproduction would hardly have been developed unless it had been of some benefit to the plants practicing it. The chief advantage seems to be in more rapid multiplication and consequently better chance to propagate the species. Only one plant is produced by each oöspore, and if this were a gametophyte with its limited number of archegonia, multiplication would be slow; but the sporophyte with its millions of spores, each capable of producing a new individual, enables the species to multiply indefinitely. On the other hand, the interposition of a gametophyte, or sexual generation, secures the introduc-

495-499. — A kind of pteridophyte (*Selaginella martensii*) with its organs of fructification : 495, a fruiting branch ; 496, a microsporophyll with a microsporangium, showing microspores through a rupture in the wall ; 497, a megasporophyll with a megasporangium ; 498, megaspores ; 499, microspores (from COULTER'S "Plant Structures").

tion of a new strain at each alternation, with the advantages of cross-fertilization (Sec. 313).

370. Microspores and Macrospores. — The method of reproduction in other pteridophytes is similar in all essentials to that of the ferns, except that in some of the orders it is even more complicated. The sporophyte, instead of producing spores which are all alike, bears two kinds of fruiting organs called *sporophylls* (spore-bearing leaves), one of which produces sporangia containing large bodies called *megaspores*, or *macrospores*, the other smaller ones called *microspores*. These large and small spores give rise to different kinds of gametophytes, one bearing archegonia, the other antheridia, and it is only by the union of a pair of gametes from each kind that an oöspore capable of producing another sporophyte can originate. This complicated arrangement may be expressed to the eye by a diagram something like the following, in which *S* stands for sporophyte, *G* for gametophyte, *mgs.* for megaspore, *mcs.* for microspore, *mgsph.* for megasporophyll, *mcsph.* for microsporophyll, *gam.* for gamete, and *oö.* for oöspore.

S. ⟶ ⎧ mgsph. ⟶ mgs. ⟶ archegonial G. ⟶ gam. ⎫ ⟶ oö. ⟶ S., etc.
　　　 ⎩ mcsph. ⟶ mcs. ⟶ antheridial G. ⟶ gam. ⎭

PRACTICAL QUESTIONS

1. Have ferns any economic use — that is, are they good for food, medicines, etc.?

2. What is their chief value?

3. Under what ecological conditions do they grow?

4. Are they often attacked by insects, or by blights and disease of any kind?

5. Of what advantage is it to ferns to have their stems under ground, in the form of rootstocks? (195.)

6. What causes the young frond of ferns to unroll? (162, 204.)

7. Name the ferns indigenous to your neighborhood.

8. Which of these are most ornamental, and to what peculiarities of structure do they owe that quality?

9. Are cultivated ferns usually raised from the spores or in some other way? Why?

STUDY OF A BRYOPHYTE

MATERIAL. — Any of the common thalloid or flat-bodied liverworts. They can generally be found growing with mosses on wet, dripping rocks and the shady banks of streams, and are easily recognized by their flat, spreading habit, which gives them the appearance of green lichens. *Marchantia polymorpha* (Fig. 500), one of the largest and best specimens for study, is common in shady, damp ground throughout the north-

500. — Umbrella liverwort (*Marchantia polymorpha*) ; portion of a thallus about natural size, showing dichotomous branching: *f, f*, archegonial or female receptacles ; *r*, rhizoids.

ern States. Lunularia, a smaller species that can be recognized by the little crescent-shaped receptacles on some of the divisions of the thallus, is abundant in greenhouses almost everywhere, on the floor, or on the sides of pots and boxes kept in damp places. Specimens of this can be procured by city classes, but the spore-bearing receptacles are seldom or never present, the species being an introduced one and possibly rendered sterile by changed conditions. *Marchantia polymorpha* is the specimen described in the text, but any allied species will do.

371. Examination of a Liverwort. — The Thallus. The broad, flat, branching organ that forms the body of the plant is the thallus. Examine the end of each branch; what do you find there? Are the two forks into which the apex of the branches divide equal or unequal? Do you see anything in these forking apexes to remind you of the heart-shaped prothallium of the fern? Are there any other points of resemblance between them? Compare the growing end with the distal one; does it proceed from a true root? Notice that as the lower end dies

501. — Under side of an archegonial receptacle enlarged. The archegonia are borne among the hairs on the under surface, which is presented to view in the figure; *f*, a spore case.

the growing branches go on increasing and reproducing the thallus.

Do you find anything like a midrib? If so, trace it along the branches and stem; where does it end? Does it seem to be formed like the midrib of a dicotyledon? Hold a piece of

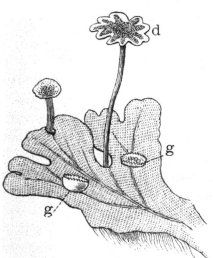

502, — Portion of a thallus bearing an antheridial disk or receptacle, *d*, and gemmæ, *g, g*.

503. — A portion of the upper epidermis of marchantia, magnified, showing rhomboidal plates with a stoma in each.

the thallus up to the light and see if you can detect any veins. Is it of the same color in all parts, and if there is a difference can you give a reason for it? Examine the upper surface with a lens. Peel off a piece of the epidermis, place it between two moistened bits of glass and hold it up to the light, keeping the upper surface toward you; what is its appearance? Observe a tiny dot near the center of the rhomboidal areas into which the epidermis is divided and compare it with your drawings of stomata (Sec. 16). What should you judge that these dots are?

372. Rhizoids. — Wash the dirt from the under side of a thallus and examine with a lens; how does it differ from the upper surface? Observe the numerous rootlike hairs, or rhizoids. What is their color? Where do they spring from? These are not true roots, but hairs that have taken upon themselves the function of absorption, and do not imply any actual differentiation of tissues.

Plant a growing thallus branch in moist earth so that the upper side will lie next the soil and watch for a week or two, noting what changes take place. What would you infer from this as to the cause of the difference between the two surfaces? Would rhizoids be of any use on the upper side? Stomata on the under side?

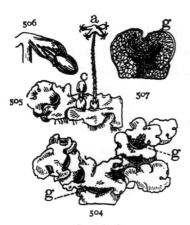

504–507. — *Lunularia*, a common liverwort : 504, portion of a thallus of about natural size : *g, g,* gemmæ; 505, a fertile plant with fruiting receptacles; 506, an enlarged section of one of the fruiting receptacles; 507, portion of a sterile thallus slightly enlarged, showing one of the crescent-shaped gemmæ from which the plant takes its name.

373. Gemmæ. — Look along the upper surface of some of your specimens for little saucer-shaped (in Lunularia, crescent-shaped) cupules or cavities. Notice the border, whether it is toothed or entire, and see if you can tell what the cupules contain. These little bodies, called *gemmæ*, are a kind of bud, by which the plant propagates itself somewhat as the onion and the tiger lily do by means of bulblets. Sow some of the gemmæ on moist sand, cover them with a tumbler to prevent evaporation, and watch them develop the thalloid structure.

374. Reproduction by Spores. — If possible, procure a thallus with upright pedicels bearing enlargements at the top like those represented in Figures 500 and 502. These are receptacles containing spore cases corresponding to the archegonia and antheridia of the fern prothallium. Notice their difference in form, the one (Fig. 502) umbrella shaped and scalloped round the edges, the other (Fig. 500) rayed, like the spokes of a wheel. The first produce antheridia only, and the second archegonia. Examine both surfaces of each, and then vertical sections, under a lens. Notice that the antheridia grow from the upper surface of the scalloped disks, the archegonia from the underside of the rayed ones, concealed in the heavy covercles that depend from the rays (Fig. 501). The archegonia and antheridia, as in the ferns, produce different kinds of reproductive cells called gametes, and so the thallus that forms the plant body of the liverwort is the gametophyte and corresponds to the prothallium of the fern. When one of the gametes from an antheridium enters an archegonium and fuses with the other kind of gamete contained there, an oöspore is formed as in the fern, which is capable of germinating and producing a new growth. But instead of falling to the ground and giving rise to an independent plant like the sporophyte of the fern, the oöspore germinates within the receptacle and produces there an insignificant spore case (*f*, Fig. 501), containing ordinary spores and thus representing in a reduced form the sporophyte that is so conspicuous a feature of the ferns. These spores, on germinating, produce the liverwort thallus body or gametophyte, thus completing the cycle of generations. Notice that in the liverwort (and all bryophytes), the thallus or gametophyte, is the important part of the plant and performs all the vegetative functions, while the sporophyte is a small, insignificant body that never becomes detached from the gametophyte and has no independent existence. In the fern and other pteridophytes just the reverse is true; the sporophyte constitutes the beautiful plant body that we all admire so much, while the gametophyte, though it does attain a separate existence, appears only as an obscure prothallium that is usually as short lived as it is inconspicuous.

375. Alternation of Generations in Seed Plants. — While the alternation of generations is more conspicuous in pteridophytes and bryophytes, it occurs also among the algæ, and is universal, though in a masked form, among the spermatophytes. It is therefore very important to have a clear idea of what it means, for the chief turning points in the life history of all plants are connected with it, and the natural relationships of the different groups and their distribution according to those relationships depend largely upon a comparison of the reproductive processes in the various classes and orders. These studies are too intricate and technical to be even outlined here; suffice it to say that some of the gymnosperms — pines, yews, cycads, etc. — show striking similarities in their reproductive processes to those of the higher pteridophytes, and through them a repetition of the most salient features of the alternation of generations in the highest seed plants has been traced. Briefly stated, we may say that the stamens of spermatophytes, and the pistils, or rather the carpels, which we saw to be transformed leaves, represent the sporophylls (Sec. 370) of the higher pteridophytes. The pollen sacs and ovules are sporangia, bearing microspores and megaspores (Sec. 370), represented respectively by the pollen grains in the anther and the embryo sac in the ovule. These go through a series of microscopic changes in the body of the ovule analogous to the production of the oöspore in the archegonia of ferns and liverworts, but the process is so obscure that to an ordinary observer the pollen grains and ovule appear to be the real gametes, and were supposed to be such, by the older botanists. The fertilized germ cell in the embryo sac (Sec. 327) corresponds to an oöspore, the endosperm found in all seeds (previous to its absorption by the cotyledons) is a rudimentary gametophyte, and the embryo in the matured seed, the undeveloped sporophyte, destined, after germination and further growth, to produce a new generation of microspores; *i.e.* pollen grains, and megaspores (embryo sac), and so on, through the cycle.

376. Relative Importance of Gametophyte and Sporophyte. — It is important to notice that the progressive diminution of the gametophyte in comparison with the sporophyte which we saw taking place in proceeding from the bryophytes to pteridophytes, reaches its climax in the spermatophytes, where it is reduced to such insignificance that it is only by certain analogies of structure and function that it can be recognized at all. It remains permanently inclosed within the walls of the ovary and is absorbed by the sporophyte during germination, or even earlier in those seeds classed as ex-albuminous. The sporophyte, on the other hand, represents the fully organized plant, and attains among dicotyledons the highest development of vegetable structure.

THE ALGÆ

MATERIAL. — Collect in a bottle some of the green scum found in stagnant pools, ditches, and sluggish streams everywhere, and variously known as frog spit, pond scum, brook silk, etc. In cities and other places where specimens are not easily procured, it can be cultivated in a simple aquarium made of a wide-mouthed glass jar with a few pebbles and sticks at the bottom.

377. Variety of Forms. — This group embraces plants of the greatest diversity of form and structure, from the minute volvox and desmids that hover near the uncertain boundaries dividing the vegetable from the animal world, to the giant kelps of the southern ocean, which sometimes attain a length of from six hundred to one thousand feet. The fresh-water algæ are all very small, and those of them that are visible to the naked eye belong mostly to the filamentous group, so called from their slender threadlike thalli, that look like bits of fine green floss floating about in the water.

378. Examination of a Specimen. — Place a drop or two of fresh pond scum on a piece of glass and examine with a lens. Of what does it appear to consist? Are the filaments all alike, or are they of different lengths and thickness? Soak a number of them in alcohol for half an hour and examine again; where has the green matter gone? Do these algæ contain chlorophyll? (Sec. 25).

379. Spirogyra. — The filamentous algæ are very numerous, and your drop of pond scum will probably contain several kinds. At least one of these, it

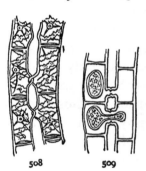

508, 509. — *Spirogyra* (magnified): 508, two filaments beginning to conjugate; 509, formation of spores.

is likely, will be a Spirogyra, as this is one of the commonest and most widely distributed of them all. This genus takes its name from the spiral bands in which the chlorophyll is usually disposed (Fig. 508) within the cells. These bands are single in some species, in others they combine and intercross in various ways, forming most beautiful patterns when viewed under the microscope. Each filament is seen, when sufficiently magnified, to consist of a number of more or less cylindrical cells joined together in a vertical row, and thus forming the simple threadlike thallus that characterizes this class of algæ. Physiologically, each cell is an independent individual, and often exists as such.

380. Vegetative Multiplication. — Some of the algæ, so far as our present knowledge goes, have only the one form of reproduction known as vegetative multiplication, or fission (splitting). A cell divides itself in two, each half grows into a distinct cell, which again divides, forming new cells, and so on, till millions of individuals may result from a single mother cell in a few days, or in some cases, in a few hours. This method of reproduction takes place in some form or other in almost all plants, the propagation by buds, tubers, rootstocks, runners, etc., among spermatophytes being nothing but a mode of vegetative multiplication.

381. Conjugation. — Another method of reproduction is by the formation of spores. In spirogyra and many other algæ the spores are formed by the method known as *conjugation*, that is, joining together. The cells of two adjacent filaments send out lateral protuberances toward each other (Fig. 509), and when the ends of these protuberances meet, the protoplasm in each contracts, the contents of one pass over into the other, the two coalesce and form a new cell but little, if any, larger than the original conjugating bodies. This cell germinates under favorable conditions and produces a new individual.

382. Diatoms and Desmids. — These two groups are alike in their microscopic size, in their simple structure, and in the interest that attaches to them on account of their enormous numbers and their great beauty and variety of form, but otherwise they are not nearly related orders. The diatoms are so different from all other vegetable structures that they are placed by some botanists in a class to themselves; others group them among the algæ. They consist of simple cells inclosed in a very hard mineral covering

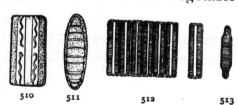

510   511   512   513

510–513. — Diatoms (highly magnified): 510, 511, *Grammatophora serpentina;* 512, 513, *Fragilaria virescens.*

formed of two valves, one of which fits over the other like the lid of a pasteboard box. They are of a brown color and of almost every conceivable shape (Figs. 510–513). Not less than ten thousand species have been described, and immense deposits of rock in various parts of the world are formed by the flinty coverings of millions of these microscopic creatures that once floated in the waters of geologic seas.

The desmids were for a long time classed with animals, but have now been handed over definitively to the botanist. They are of a bright green color, and are further distinguished from the diatoms by their perfect bilateral sym-

metry ; that is, both sides of a cell are just alike. They
are found only in fresh water ; diatoms inhabit either salt
water or fresh.

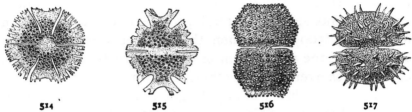

514   515   516   517

514–517.—Desmids (highly magnified) : 514, *Micrasterias papillifera* ; 515, *Micrasterias morsa* ; 516, *Cosmarium polygonum* ; 517, *Xanthidium aculeatum.*

383. Place in Nature. — Algæ exist in vast multitudes
both as to the number of species and of individuals. They
all contain chlorophyll, but in a few fresh-water forms
and in most seaweeds it is obscured by pigments of brown
or red to which the brilliant coloration of these plants is
due. The presence of these pigments probably has some
relation to their peculiar environment, especially in the
case of those growing in deep water, where the action of
light upon the chlorophyll is greatly diminished and altered
by refraction. Their variations in color form a convenient
basis of classification, and botanists divide algæ into six
great orders, according to their color. The spirogyra and
most fresh-water species belong to the order of *Chloro-
phyceæ*, or Green Algæ. This class is of special interest
because from it all the higher forms of vegetable life are
believed to have been derived.

PRACTICAL QUESTIONS

1. Are any of the green algæ parasitic ? How do you know ? (25.)

2. What is their effect upon the atmosphere ; that is, do they tend
to purify it by giving off oxygen, or the reverse ? (24, 25.)

3. Why is their presence in water regarded as denoting unhygienic
conditions ?

4. Refer to the experiment in Section 22, and account for the bubbles and froth that usually accompany these plants in the water.

5. Can you suggest any other causes than the elimination of oxygen
that might produce the same effect ?

6. Is the presence of these gas bubbles of any use to the plants?

XI. FUNGI

384. What is a Fungus? — The fungi are all (with a few doubtful exceptions) parasites or saprophytes which have lost their chlorophyll and become incapable of supporting an independent existence. Biologists are divided as to their position in the genealogical tree of life. The weight of authority at present seems to incline to the view that they are degenerate forms derived from the algæ, while others regard them not as degraded descendants of higher forms, but as representatives of the lowest primordial types from which higher organizations have arisen. If they represent a degraded and degenerate type, they have been so modified by their parasitic habits as greatly to obscure their relationship and render their position in the general scheme of life a very doubtful one. They represent an offshoot, or side branch as it were, of the great evolutionary line, and so will be considered in a chapter by themselves.

385. Economic Importance. — On account of their immense numbers, reaching at present the enormous total of forty-five thousand known species, and of the parasitic habit, which causes them to enter the bodies of other plants and of animals, fungi are of great economic importance, especially the various microscopic forms grouped under the head of Bacteria. By their rapid multiplication within the blood and the tissues of their victims they produce the most fatal and destructive diseases. They are the smallest living organisms, and are always floating in the atmosphere, so that with every breath we draw, large numbers of them are inhaled. Fortunately, however, most of them are harmless, unless inhaled in very great numbers or

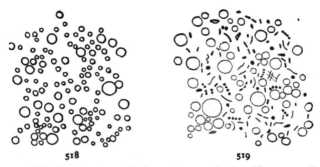

518, 519.— Milk (highly magnified): 518, pure, fresh milk; 519, milk that has stood for hours in a warm room in a dirty dish, showing fat globules and many forms of bacteria.

under certain unhealthful conditions, while a few, such as the yeast fungus and the bacteria concerned in the processes of decomposition, are very useful. The presence of

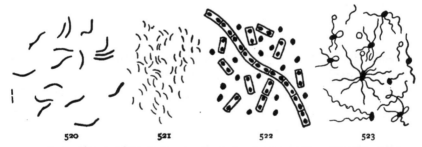

520–523.— Forms of bacteria: 520, bacteria of consumption (*Bacillus tuberculosis*); 521, cholera bacillus; 522, bacilli of anthrax, showing spores; 523, typhoid bacillus.

bacteria in the soil is also of importance sometimes, since through their agency the nitrogen compounds are rendered soluble by the roots of plants (Sec. 198).

386. Difficulty of Classification. — The life history of fungi in general is very obscure and difficult to trace, both on account of the microscopic size of the great majority of them, and of the curious habit of polymorphism exhibited by many species; that is, the same individual appears under entirely different forms at different stages of its existence, like an insect undergoing metamorphosis, so that it is often impossible to tell whether a given specimen belongs to a distinct group or is merely a form of the same species at a different stage of its existence.

Our knowledge of them being so imperfect, their classification is in great confusion, and any grouping of them must be considered as in a great measure provisional only.

PRACTICAL QUESTIONS

1. Why ought preserved fruits and vegetables to be scalding hot when put into the can? (385.)
2. Why is it necessary to exclude the air from them? (385.)
3. Why does using boiled water for drinking render a person less liable to disease? (142, 385.)

MUSHROOMS

MATERIAL. — Any kind of gilled mushroom in different stages of development, with a portion of the substratum on which it grows, containing some of the so-called spawn. In city schools the common mushroom sold in the markets (*Agaricus campestris*) can usually be obtained without difficulty. It would be advisable to buy some of the spawn and raise a crop in the schoolroom, as then all parts of the plant would be on hand for examination. Full directions for cultivating this fungus are given in Bulletin 53 of the U.S. Department of Agriculture. From six to twelve hours before the lesson is to begin, cut the stem from the cap of a mature specimen close up to the gills, lay the gills downward on a piece of clean paper, cover them with a bowl or pan to keep the spores from being blown about by the wind and leave them until a print (Fig. 532) has been formed.

524. — Deadly agaric (*Amanita phalloides*), showing the broad pendent annulus, *a*, formed by the ruptured veil, the cup at the base, *c*, and floccose patches on the pileus, left by the breaking up of the volva.

387. Examination of a Typical Specimen. — The most highly specialized of the fungi, and the easiest to observe on account of their size and abundance, are the mushrooms that are such familiar objects after every summer shower. The gilled kind — those with the rayed laminæ under the cap — are usually the most easily obtained. Gather a specimen of some of these according to the directions given above, and examine them as soon as possible, since they decay very quickly.

388. The Mycelium. — Examine some of the white fibrous substance usually called spawn, through a lens. Notice that it is made up of fine white threads interlacing with each other, and often forming webby mats that ramify to a considerable distance through the substratum of rotten wood or other material upon which the fungus grows. These threads are called *hyphæ*, and are apt to be mistaken for roots, but they are really the thallus or true vegetative body of the plant, the part rising above ground and usually regarded as the mushroom, being only the fruit, or reproductive organ. The thallus of all fungi is called a *mycelium* from *mycetes*, a Greek word meaning fungi.

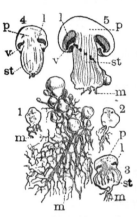

525. — Mycelium of a mushroom (*Agaricus campestris*) with young buttons (fruiting organs) in different stages: 1, 2, 3, 4, 5, sections of fructification at successive periods of development; *m*, mycelium; *st*, stipe; *p*, pileus; *l*, gill, or lamina; *v*, veil.

389. The Button. — Look on the mycelium for one of the small round bodies called buttons (Fig. 525). These are the beginning of the fruiting body, popularly known as the mushroom, and are of various sizes, some of the youngest being barely visible to the naked eye. After a time they begin to elongate and make their way out of the substratum.

390. The Veil and Volva. — Make a vertical section through the center of one of the larger buttons after it is well above ground, and sketch. Notice whether it is entirely enveloped from root to cap in a covering membrane — the *volva* (Fig. 526, *a*) — or whether the enveloping membrane extends only from the upper part of the stem to the margin of the cap — the *veil* (Fig. 526, *d*); whether it has both veil and volva, or finally whether it is naked, that is, devoid of both.

Next take a fully expanded specimen and observe

391. The Stipe, or stalk. Notice as to length, thickness, color, and position, that is, whether it is inserted in the center of the cap or to one side (excentric), or on one edge (lateral). Observe the base, whether bulbous, tapering, or straight, and whether surrounded by a cup, or merely by concentric rings or ragged bits of membrane (the remains of the volva).

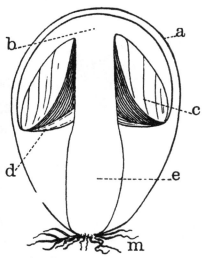

526. — Diagram of unexpanded *Amanita*, showing parts: *a*, volva; *b*, pileus; *c*, gills; *d*, veil; *e*, stipe; *m*, mycelium.

Look for the annulus or ring (remains of the veil) near the insertion of the stipe into the cap, and if there is one, notice whether it adheres to the stipe, or moves freely up and down as in Figure 527, *a*; whether it is thick and firm, or broad and membranous so that it hangs like a sort of curtain round the upper part of the stipe (Fig. 524, *a*). Break the stem and notice whether it is hollow or solid, observe also the texture, whether brittle, cartilaginous, fibrous, fleshy, etc. Next observe the

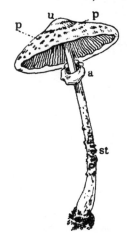

527. — Parasol mushroom (*Lepiota procera*), showing movable annulus: *st*, stipe; *a*, annulus, or ring; *u*, umbo; *p, p*, floccose patches left by volva.

392. Pileus, or cap, as to color and surface, whether dry, or moist and sticky; smooth, or covered with scurf or scales left by the remains of the volva, as it was stretched and broken up by the expanding cap (Fig. 527, *p, p*). Note also the size and shape, whether conical, expanded, funnel shaped (infundibuliform, Fig. 528); umbonate, having a protuberance at the apex (Fig. 527), etc; whether the margin is turned up at the edge (revolute, Fig. 524), or under (involute, Fig. 527). Look at the under surface and examine

528. — Chanterelle (*Cantharellus cibarius*), with infundibuliform pileus and decurrent gills.

393. The Gills, or laminæ. —

Notice whether they are •broad or narrow, whether they extend straight from stem to margin or are rounded at the ends, or are curved, toothed, or lobed in any way. Notice their attachment to the stipe, whether free, not touching it at all; adnate, attached squarely to the stem at their anterior ends; or decurrent, running down upon the stem for a greater or less distance (Fig. 528).

394. The Hymenium. — Cut

a tangential section through one side of the pileus and sketch as it appears under the lens. If a very thin cross section of one of the gills is made and placed under the microscope it will appear as in Figure 529. More highly magnified sections are shown in Figures 530, 531. The blade of the gill, called the *trama*, is covered on both sides by a membranous layer bearing elongated club-shaped cells set

529–531. — Sections of a gilled mushroom: 529, through one side, showing sections of the pendent gills, *g*, *g*, (slightly magnified); 530, one of the gills more enlarged, showing the central tissue of the trama, *tr*, and the broad border formed by the hymenium, *h*; 531, a small section of one side of a gill very much enlarged, showing the club-shaped basidia, *b*, *b*, standing at right angles to the surface, bearing each two small branches with a spore, *s*, *s*, at the end. The sterile paraphyses, *p*, are seen mixed with the basidia.

upon it at right angles to the surface (Fig. 530). Some of these put out from two to four, or in some species as many as eight little prongs, each bearing a spore (Fig. 531, *s*, *s*), while others remain sterile. The spore-bearing cells are called *basidia*, the sterile ones, *paraphyses*, and the whole spore-bearing surface together, the *hymenium*, from a Greek word meaning a membrane. It is from the presence of this expanded fruiting membrane that the class of mushrooms we are considering gets its botanical name, *Hymenomycetes*, membrane fungi.

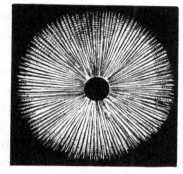

532. — Spore print of a gilled mushroom.

395. Spore Prints. — When the gills are ripe they shed their spores in great abundance. Take up the pileus that was laid on paper as directed under *Material*, on page 273, and examine the print made by the discharged spores ; it will be found to give an exact representation of the under side of the pileus.

The hymenium is not always borne on gills, but is arranged in various ways which serve as a convenient basis for distinguishing the different orders. In the *Polyporei*, to which the edible boletus belongs (Figs. 533, 534), the basidia are placed along the inside of little tubes that line the under side of the pileus, giving it the appearance of a honeycomb. In another order,

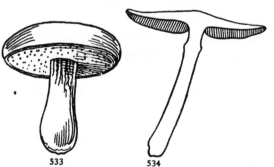

533, 534. — A tube fungus (*Boletus edulis*) : 533, entire ; 534, section, showing position of the tubes.

the porcupine fungi, they are arranged on the outside of projecting spines or teeth, while in the morelles they are held in little cups or basins.

396. The Spores. — Notice the color of the spores as shown in the spore print. This is a matter of importance in distinguishing gill-bearing fungi, which are divided into five sections according to the color of the spores. One source of danger, at least, to mushroom eaters would be avoided if this difference was always attended to, for the deadly amanita (*A. phalloides*), and the almost equally dangerous fly mushroom (*A. muscaria*), both have white spores, while the favorite edible kind (*Agaricus campestris*), though white gilled when young, produces dark, purple-brown spores that can not fail to distinguish it clearly for any one who will take the trouble to make a print.

535. — Diagram of a gilled mushroom.

Sketch a longitudinal section through the center of a well-developed mushroom, as shown in Figure 535, labeling the different parts that you can distinguish, and bringing out as well as you can the points observed in your examination of the living specimen.

397. Mushrooms and Toadstools. — The popular distinction which limits the term "mushroom" to a single species, the *Agaricus campestris*, and classes all others as toadstools, has no sanction in botany. All mushrooms are toadstools and all toadstools are mushrooms, whether poisonous or edible. The real distinction is between mushrooms and puff balls, the former term being more properly applied only to that class of fungi which have the hymenium ,or spore-bearing surface exposed.

398. Food Value. — The food value of mushrooms has been greatly exaggerated. They contain a large proportion of water, often over ninety per cent, and the most valued of them, the *Agaricus campestris*, bears a very close resemblance to cabbage in its nutrient properties. They are pleasant relishes, however, and as agreeable articles of diet, are not to be despised.

1. Why are mushrooms generally grown in cellars? (24, 384.)

2. Name any fungi you know of that are good for food or medicine or any other purpose.

3. Name the most dangerous ones you know of.

4. Do you find fungi most abundant on young and healthy trees, or on old, decrepit ones? Account for the difference. (384.)

5. Do you ever find them growing upon perfectly sound wood anywhere?

6. Is it wise to leave old, unhealthy trees and decaying trunks in a timber lot?

RUSTS

MATERIAL. — A leaf of wheat affected with red rust. A leaf or a stalk with black rust. Some barberry leaves with yellowish pustules on the under side that look under the lens like clusters of minute white corollas (see Fig. 542). As the spots on barberry occur in spring, the red rust in summer, and the black rust in autumn, the specimens will have to be gathered as they can be found, and preserved for use.

In the southern States barberry occurs but rarely or not at all, and a different species of rust, the orange leaf (*Puccinia rubigo-vera*), is more common than the ordinary wheat rust (*Puccinia graminis*), but the two are so much alike that the directions given will do for either. If the orange leaf rust is used, the cups and pustules should be looked for on plants of the borrage family — comfrey, hound's-tongue, etc. Leaves of oats or other infected grasses may be used, but wheat is to be preferred, as the life history of the common wheat rust (*P. Graminis*) has been more clearly traced than that of any other variety. The apple scab fungus may be used instead of wheat if more convenient. In this case, provide apple or haw leaves affected with scab, and some of the common excrescences known as cedar apples.

399. Red Rust.— Uredo Stage. Examine a leaf of " red rusted " wheat under the lens, and notice the little oblong brown dots that cover it. These are the sori, or clusters of sporangia that have formed upon the surface. Viewed under the microscope the red rust is seen to consist of a mycelium that ramifies through the tissues of the leaf and bears clusters of single-celled reddish spores that break through the epidermis and form the reddish brown spots and streaks from which the disease takes its name. These spores, falling upon other leaves, germinate in a few hours and form new mycelia, from which, in six to ten days, fresh spores arise. Formerly this was thought to complete the life history of the fungus, to which the name of *Uredo* was given. It is now known, however, that the red rust is merely a stage in the life cycle of the plant, and to this stage the old name uredo is applied, and the spores are called uredospores.

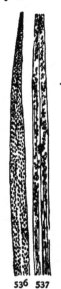

536 537

536, 537. — Leaf of wheat affected with orange leaf rust, *Puccinia rubigo-vera*, uredo stage: 536, upper side of leaf; 537, under side.

400. Black Rust. — Next examine with your lens a part of the plant attacked by black rust. Do you observe any difference except in the color? Do the two kinds of rust attack all parts of the plant equally? If not, what part does each seem to affect more particularly? At what season does the black rust appear most abundantly?

It was formerly supposed that black rust was caused by a different fungus from that producing red rust, and to it the name *Puccinia* was given, but it is now known to be only another phase of the same parasite that produces the red rust. The name "Puccinia" is retained as a general designation for all fungi undergoing these two phases, and the particular form of fungus that we are now considering is known in all its stages as *Puccinia graminis*.

538. — Stalk with *Puccinia graminis*, teleuto stage.

401. Teleutospores. — Toward the end of summer the same mycelium that bore the uredospores begins to develop the dark spore clusters that give to black rust its characteristic color and its name. After this the uredospores soon cease to be developed at all, and only the dark ones called teleutospores are produced. These remain on the culms in the stubble fields over winter, ready to begin the work of reproduction in spring, whence they are called "winter spores," in contradistinction to the uredos or "summer spores," whose activity seems to be confined to the warm months.

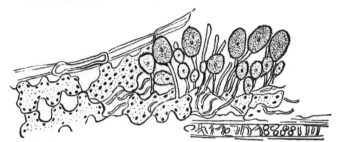

539.—Uredospores of wheat rust, *Puccinia graminis,* magnified (from COULTER'S " Plant Structures ").

Under the microscope the teleutospores appear as long, two-celled bodies with very thick black walls (Fig. 540). Since they are developed from the same mycelium with the

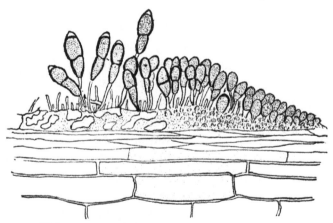

540.—Teleutospores of wheat rust, magnified (from COULTER'S " Plant Structures ").

uredospores, and are not a product of the latter, but collateral with them, the two constitute a single generation, and belong to one and the same stage in the life history of the plant.

402. Sporidia. — In spring the teleutospores begin to germinate, each cell producing a small filament, from which arise in turn several small branches. Upon the tip of each of these branches is developed a tiny sporelike body called a *sporidium* (Fig. 541), which continues the generation of the rust fungus through the next stage of its existence. The filament which bears these sporidia is not parasitic, but when the sporidia ripen and the spores contained in them are scattered by the wind, there begins a second parasitic phase, which forms the most curious part of this strange life history.

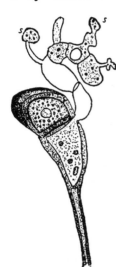

541. — Teleutospore germinating and forming sporidia, *s, s*, (from COULTER'S "Plant Structures").

403. The Æcidium. — Examine now the under side of your barberry leaves (or comfrey, etc., if red rust is used), for clusters of small whitish bodies that appear under the lens like little white corollas with yellow anthers in the center. More highly magnified, this yellow substance is seen to be composed of regular layers of colored spores. The corolla-like receptacles containing them, popularly known as "cluster cups," are borne on a mycelium produced from the spores described in the last paragraph. This mycelium is parasitic on

542. — Cluster cups of apple rust (*Rostelia*), the æcidium stage of the "cedar apple" fungus.

barberry or other leaves, according to the kind of fungus, and was long believed to be a distinct plant, to which the name *Æcidium* was given. This term (pl. *Æcidia*) is now applied to the cluster cups, and those fungi which at any period of their life history produce them are called *Æcidiomycetes*, Æcidium fungi.

404. Connection between Barberry and Wheat Rust. — There had long existed a popular belief, both in this country and in England, that the presence of barberry bushes near grain fields produced rust, or mildew, as it is called in England. There is a village in Norfolk that long went by the name of "Mildew Rollesby," on account of the mildewed grain caused, it was believed, by the abundance of barberry bushes in the neighborhood. These were cut down and mildew at once disappeared. Repeated instances of the kind led a few men of science to suspect that the popular belief might be something more than a mere superstition, after all. Experiments were made which showed that grain planted in the vicinity of a barberry bush infected with æcidia developed rust immediately after the æcidia spores matured, and that rust was most abundant in the direction in which the wind carried the spores. Further experiment showed that æcidia spores would not germinate directly on barberry; in other words, æcidia would not reproduce æcidia directly, but only after passing through one or more intermediate stages, and thus it was proved beyond a doubt that these fungi are not independent plants, but merely a phase in the life history of the Puccinia.

405. The Life Cycle. — Taking the first phase of the season as our starting point, the life cycle of the wheat rust consists of three stages carried on by four different kinds of spores: (1) The non-parasitic stage, which originates from teleutospores, and produces sporidia; (2) The æcidium phase, which arises from the sporidia, is parasitic on barberry, and produces spores that germinate on grain; (3) The uredo-teleuto phase, parasitic on grain. The first, or sporidia stage, which is too small to be discoverable except by the microscope, escapes the notice of the ordinary observer, and the third, producing two kinds of spores, uredo and teleuto, has the appearance of being two separate stages, so that to one unacquainted with the facts, the life cycle would seem to consist of a red rust or uredo stage, a black rust, or teleuto stage, and an æcidium stage. The last is often omitted. In many cases, as in our own southern States, where there are no barberries to act as hosts, the sporidia germinate directly upon young wheat, without passing through the cluster cup stage, and the orange leaf

rust is known to be capable of propagating year after year
in the uredo stage alone,[1] the spores surviving through the
winter on volunteer grains and other grasses.

406. Cedar Apples. — An excellent subject for study is
the common fungus (*Gymnosporangium*) that produces
upon red cedar twigs the large excrescences familiarly
known as "cedar apples." It is related to the wheat rusts,
but has only two phases, its spores germinating and pro-

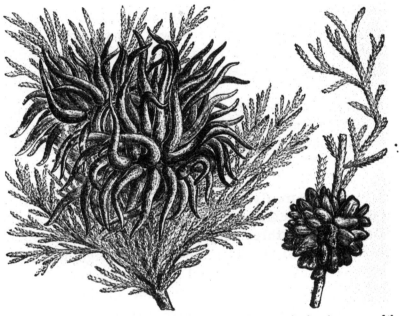

543. — Two species of "cedar apple" (*Gymnosporangium*), showing stage of the
apple rust fungus corresponding to the uredo-teleuto stages of wheat rust (from
COULTER'S "Plant Structures").

ducing æcidia upon the leaves of apple, hawthorn, and
other kindred plants. In this stage it is known as *Rostelia*,
and is the cause of apple rust and other similar orchard
diseases. Specimens are generally easy to obtain and can
be studied by the same methods outlined in the foregoing
paragraphs.

407. Polymorphism. — Plants that pass through different
stages in their life history are said to be *polymorphic*, that

[1] Bulletin 16, United States Department of Agriculture.

is, of many forms. The habit is very common among the lower forms of vegetation. The fact that one or more of the phases are sometimes omitted, as the æcidium phase of wheat rust in warm climates, suggests the idea that it may be of use in helping the plant to tide over difficult conditions. Blackberry, anemone, groundsel, buckthorn, and many other common plants are known to harbor æcidia, but what particular species of uredo and puccinia and æcidium belong together in any one case, it is impossible to determine without continued observation and experiment.

408. Difference between Polymorphism and Alternation of Generations. — These two processes must not be confounded. A polymorphic plant, so far as we know, may reproduce itself indefinitely by means of simple spores without the intervention of gametes and oöspores, but to constitute an alternation of generations there must intervene somewhere in the life history the union of two unlike spores (gametes) to form an oöspore, with the alternate appearance of sporophyte and gametophyte.

PRACTICAL QUESTIONS

1. Is a farmer wise to leave scabby and mildewed weeds and bushes in the neighborhood of his grain fields? (407.)

2. Are there any objections to the presence of volunteer grain stalks along roadsides and in fence corners during winter? (405.)

3. Should cedar trees be allowed to grow near an apple orchard? Give a reason for your answer. (406.)

4. Should diseased plants be plowed under? (402.)

5. What disposition should be made of them?

6. Ought diseased fruits be left hanging on the tree?

7. Why is it necessary to pick over and discard from a crate or bin all decaying fruits and vegetables?

FIELD WORK

The study of fungi can be carried on only to a very limited extent without the use of a compound microscope, and all serious work of the kind must be conducted in the laboratory. The general observer, however, may do some practical work by learning to recognize the various blights, rusts, mildews, etc., by their effects upon the vegetation of his neighborhood.   Learn to know at a glance whether a given field or or- ' chard is suffering from leaf curl, scab, the yellows, bunt, smut, mildew, etc.

A systematic study of mushrooms will be found very interesting from both a scientific and a dietetic point of view for those who have leisure to undertake it and means to expend on the rather costly literature that deals with the subject.

SYSTEMATIC BOTANY

409. Now that some knowledge has been obtained of the structure of plants, their analysis and classification can be taken up with both profit and pleasure.   To know the place of a species in the great scheme of life, and understand what is to be expected of it in its normal family relations is necessary before we can appreciate justly its adaptations to the surrounding conditions in its struggle for existence.   It is not advisable to spend too much time in the mere identification of species, but enough should be examined and described to familiarize the student with the distinctive characteristics of the principal botanical groups.

410. Botanical Terminology is in a very unsettled state at present, owing to disagreements among botanists as to the use of certain terms, but this does not affect the general principles of classification and nomenclature.   All the known plants in the world, varied and multitudinous as they are, numbering not less than one hundred and twenty thousand species of the seed-bearing kind alone, are ranged according to certain resemblances of structure, into a number of great groups known as families or orders.   The names of these families are distinguished by the ending *aceæ ;* the rose family, for instance, are the *Rosaceæ ;* the pink family, *Caryophyllaceæ ;* the walnut family, *Juglandaceæ*, etc.

411. Genera and Species. — Each of these families is divided into lesser groups called *genera* (singular, *genus*), characterized by similarities showing a still greater degree of affinity than that which marks the larger groups or orders ; and finally, when the differences between the individual plants of a kind are so small as to be disregarded, they are considered to form one species, just as all the common morning-glories, of whatever shade or color, belong to the species *Ipomea purpurea*. The small differences that arise within a species as to the color and size of flowers, and other minor points, constitute mere varieties and have no special names applied to them. The line between varieties and species is not clearly defined, and in the nature of things can never be, since progressive development, through slow but unceasing change, is the law of all life.

In botanical descriptions the name both of the species and of the genus is given, just as in designating a person, like Mary Jones or John Robinson, we give both the surname and the Christian name. The genus, or generic name, answers to the surname, and that of the species to the Christian name — except that in botanical nomenclature the order is reversed, the generic, or surname coming first, and the specific or individual name last; for example, *Ipomea* is the generic, or, surname, of the morning-glories, and *purpurea* the specific one.

412. How to use the Key. — Any good manual will do; Gray's "School and Field Book" is perhaps the best available at present for the States east of the Mississippi. A little reference to what has already been said on the subject of classification in Sections 126–129, will make its use clear. Suppose we want to find out to what botanical species the morning-glory, or the sweet potato, for instance, belongs. Turning to the key we find the sub-kingdom of Phænerogams — flowering, or seed-bearing plants — divided into two great classes, Angiosperms and Gymnosperms, as already explained in the Sections referred to. A glance will show that our specimen belongs to the former class. Angiosperms, again, are divided into the two subclasses of Dicotyledons and Monocotyledons (Sec. 129). We at once recognize our plant, by its net-veined leaves and pentamerous flowers as a dicotyledon (Secs. 37, 302), and turning again to the key, we find this subclass divided into three great groups: Sympetalous (called also Monopetalous and Gamopetalous); Apopetalous (or Polypetalous); and Apetalous. A glance will refer our blossom to the sympetalous or monopetalous group, which we find divided into two sections, characterized by the superior or inferior ovary (Secs. 289, 294). A little examination will show that the morning-glory belongs to the former class, which is in turn divided into two sections, according as the corolla is *regular*, or *more or less irregular*. We see at once that we must look for our specimen in the former class. This we find again subdivided into four sections according to the number and position of the stamens, and we find that the morning-glory falls under the last of these; "Stamens as many as the lobes or parts of the corolla and alternate with them." A very little further search brings us to the family *Convolvulaceæ*, and turning to that title in the descriptive analysis, page 306, we find under the genus, *Ipomea*, a full description of the common morning-glory, in the species *Ipomea purpurea*, and of the sweet potato in the species *Ipomea batatas*.

APPENDIX

BOOKS FOR READING AND REFERENCE

An excellent bibliography, accompanied by short explanatory notices of the works most useful to teachers of botany, will be found in the seventh chapter of Ganong's Teaching Botanist, which the reader is advised to consult. Some of the books mentioned there, however, are too technical to fall within the scope of this work, and others of value have appeared since the list was compiled. The references in the following pages have been arranged, as far as possible, with regard to the subjects treated in the different chapters of the present work; but the order of treatment by different authors varies so, that it has been impossible to specialize closely. Some of the references given under one head will be found to contain matter equally applicable to other subjects, and what is suitable for one section of a chapter will perhaps have no connection with the other parts of the same chapter. The most that can be done is to furnish a list for general guidance, as an aid to those teachers who have not access to well-stocked libraries.

The price of all the works named has been given wherever it could be ascertained, and also the address of the publishers and date of publication. Where more than one reference is made to the same work, these data are omitted after the first. With one or two exceptions, no foreign publications, unless reprinted in this country, are included in the list. Nearly all the articles quoted from the Year Books of the Department of Agriculture, and other government publications, have been reprinted in pamphlet form, and can be obtained free by addressing the Bureau of Publication, United States Department of Agriculture, Washington, D.C. A circular containing a list of all the publications of the department will be sent free on application.

CHAPTER II

Allen: Story of the Plants. Chaps. IV and V. D. Appleton & Company, N.Y. 35 cents.

Darwin: Insectivorous Plants. D. Appleton & Company. 1886. $2.00.

Gray: Structural Botany, pp. 85–131. American Book Company, N.Y. 1880. $2.00.

Goodale: Physiological Botany, pp. 337–353 and 409–424. American Book Company. 1885. $2.00.

Leavitt: Outlines of Botany, pp. 66–98. American Book Company. 1901. $1.00.

Lubbock: Flowers, Fruits, and Leaves; Last Part. Macmillan Company, N.Y. 1884. $1.25.

Ruskin: Modern Painters. Vol. V, Chaps. I, II, IV, V, IX, X. John Wiley & Sons, N.Y.

Dana: Plants and Their Children, pp. 135–185. American Book Company. 1896. 65 cents. (An elementary work, but full of excellent suggestions and examples.)

Thoreau: Autumn Tints, from "Excursions in Field and Forest." Houghton, Mifflin & Company, Boston. 1891. $2.00.

Treat: Home Studies in Nature. Part III. American Book Company. 90 cents.

Ward: Disease in Plants. Chaps. III and IV. Macmillan Company. 1901. $1.60.

Report of the Division of Forestry: United States Department of Agriculture. 1899.

CHAPTER III

Bailey: The Evolution of Our Native Fruits. Macmillan Company. 1898. $2.00.

Gray: Structural Botany. Chap. VII.

Leavitt: Outlines of Botany, pp. 147–156.

Lubbock: Flowers, Fruits, and Leaves. Part II.

Thoreau: " The Succession of Forest Trees " and " Wild Apples," from " Excursions in Field and Forest."

Dana: Plants and Their Children, pp. 27–49.

CHAPTER IV

Dana: Plants and Their Children, pp. 50–98.

Goodale: Physiological Botany, pp. 205 and 384–396.

Leavitt: Outlines of Botany, pp. 7–23.

Lubbock: Seeds and Seedlings. D. Appleton & Company. 1892. 4 vols. $10.00.

Year Book of the United States Department of Agriculture. 1894. Pure Seed Investigation, pp. 389–408 ; Water as a Factor in the Growth of Plants, pp. 165–176.

Year Book. 1895. Oil-producing Seeds, pp. 185–204 ; Testing Seeds at Home, pp. 175–184.

Year Book. 1896. Migration of Weeds, pp. 263–286 ; Superior Value of Large, Heavy Seed, pp. 305–322.

Year Book. 1897. Additional Notes on Seed Testing, pp. 441–452.
Year Book. 1898. Improvement of Plants by Selection, pp. 355–376.
Grass Seed and its Impurities, pp. 473–494.

CHAPTER V

Gray: Structural Botany, pp. 27–39 and 56–64.
Leavitt: Outlines of Botany, pp. 34–45; 58–60.
Ward: Disease in Plants. Chaps. V, VI, and VII.
Year Book of the United States Department of Agriculture. 1894.
Grasses as Sand and Soil Binders, pp. 421–436.

CHAPTER VI

Apgar: Trees of the Northern United States. Chaps. II, V, and VI.
American Book Company. 1892. 55 cents.
Leavitt: Outlines of Botany, pp. 45–56; 212–226; 229–240.
Pinchot: A Primer of Forestry. Bulletin No. 24. Division of Forestry: United States Department of Agriculture. 1899.
Popular Science Monthly, September, 1901. Plants as Water Carriers.
Popular Science Monthly, March, 1902. The Palm Trees of Brazil.
Ward: The Oak. D. Appleton & Company. 1892. $1.00.
Ward: Disease in Plants. Chaps. XXI, XXII, XXVI, XXIX.
Ward: Timber and Some of its Diseases. The Macmillan Company.
1889. $1.75.
Year Book. 1894. Forestry for Farmers, pp. 461–500. (Bulletin 67.)
Year Book. 1895. Principles of Pruning and Care of Wounds in
Woody Plants, pp. 257–268.
Year Book. 1898. Pruning of Trees and Other Plants, pp. 151–166.

CHAPTER VII

Gray: Structural Botany. Chap. V.
Huntington: Studies of Trees in Winter. Knight & Millet, Boston.
1900. $2.25.
Leavitt: Outlines of Botany, pp. 23–33 and 138–143.
Lubbock: Buds and Stipules. D. Appleton & Company. $1.25.
Ruskin: Modern Painters. Chaps. III, VI, and VII.

CHAPTER VIII

Allen: Flowers and Their Pedigrees. D. Appleton & Company. $1.50.
Dana: Plants and Their Children, pp. 187–255.
Darwin: Different Forms of Flowers of the same Species. D. Appleton
& Company. $1.50.

Darwin: On the Fertilization of Orchids. $1.75.

Darwin: Cross- and Self-fertilization in the Vegetable Kingdom. Chaps. I and II. $2.00. Both by D. Appleton & Company. 1886.

Gray: Structural Botany, pp. 163–214; 215–242.

Henslow: The Origin of Floral Structures through Insects and Other Agencies. D. Appleton & Company. 1895. $1.75.

Leavitt: Outlines of Botany, pp. 99–138.

Lubbock: Flowers, Fruits, and Leaves; First Part.

Lubbock: British Wild Flowers in Relation to Insects. Macmillan Company. 1893. $1.25.

Müeller: The Fertilization of Flowers. Macmillan Company. 1893. 21s. (about $5.00).

Trelease: The Yucca Moth and Yucca Pollination. (Report of the Missouri Botanical Garden.) 1892.

Ward: Disease in Plants. Chap. VIII.

Year Book. 1896. Seed Production and Seed Saving, pp. 207–216.

Year Book. 1897. Hybrids and Their Utilization in Plant Breeding, pp. 383–420.

Year Book. 1898. Pollination of Pomaceous Fruits, pp. 167–180.

Year Book. 1899. Progress of Plant Breeding in the United States, pp. 465–490.

Year Book. 1900. Smyrna Fig Culture in the United States, pp. 79–106.

CHAPTER IX

Allen: Colin Clout's Calendar. Particularly Chaps. XXXVI–XXXVIII. Funk & Wagnalls Company, N.Y. 1883. Cloth, $1.00; paper, 25 cents.

Bailey: The Survival of the Unlike. Macmillan Company. 1897. $2.00.

Darwin: The Variation of Animals and Plants under Domestication. Chaps. IX–XII. D. Appleton & Company. 2 vols. $5.00.

Dawson: The Geological History of Plants. D. Appleton & Company. $1.75.

Ward: Disease in Plants. Chaps. VII, X, XI, XVII, and XIX.

Contributions from the United States National Herbarium: —

The Plant Covering of Ocracoke Island. Thomas Kearney. Vol. V, No. 5. 1900.

Plant Life of Alabama. ´Chas. Mohr. Vol. VI. 1901.

Year Book. 1894. The Geographic Distribution of Animals and Plants in North America, pp. 203–214.

Year Book. 1895. The Grasses of Salt Marshes, pp. 325–332.

Year Book. 1898. Weeds in Cities and Towns, pp. 193–200. Forage Plants on Alkali Soils, pp. 535–550.

The Water Hyacinth in its Relation to Navigation in Florida. Bulletin 18. United States Department of Agriculture.

CHAPTER X

Clute: Our Ferns in Their Native Haunts. Frederick A. Stokes & Company, N.Y. 1901. $2.15.

Huxley and Martin: "Algæ" and "A Study of Pteris Aquilina," from "A Course of Elementary Instruction in Practical Biology." Macmillan Company. 1886. $2.60.

Leavitt: Outlines of Botany, pp. 163–183, 198–212.

Parsons (Mrs. Dana): How to know the Ferns. Charles Scribner's Sons, N.Y. 1899. $1.50.

Underwood: Our Native Ferns and Their Allies. 6th revised edition. Henry Holt & Company, N.Y. 1900. $1.00.

CHAPTER XI

Atkinson: Mushrooms: Edible and Poisonous. Andrus & Church, Ithaca, N.Y. 1900. $3.00.

Gibson: Our Edible Toadstools and Mushrooms. Harper & Brothers, N.Y. $7.50.

Leavitt: Outlines of Botany, pp. 183–197.

Marshall: The Mushroom Book. Doubleday, Page & Company, N.Y. 1901. $3.00.

Massee: Text Book of Plant Diseases. Macmillan Company, N.Y. 1899. $1.60.

McIlvaine: One Thousand American Fungi. 2d edition. The Bowen Merrill Company, Indianapolis. 1902. $5.00.

Ward: Timber and Some of its Diseases. Chaps. V, VI, VII, X–XIII.

Year Book. 1894. Grain Smuts: Their Cause and Prevention, pp. 409–420.

Report of the Department of Agriculture. 1885. Twelve Edible Mushrooms of the United States. (Reprinted as a Bulletin, 1890.)

Report of the Secretary of Agriculture. 1890. Mushrooms of the United States, pp. 366–373.

Year Book. 1897. Some Edible and Poisonous Fungi, pp. 453–470.

Year Book. 1900. Fungous Diseases of Forest Trees, pp. 199–210.

Cereal Rusts of the United States. Bulletin No. 16. United States Department of Agriculture.

How to grow Mushrooms. Bulletin 53.

Mushroom Poisoning. Circular No. 13. Department of Agriculture.

BOOKS FOR GENERAL REFERENCE

1. Allen: The Story of the Plants. D. Appleton & Company, N.Y. 35 cents.
2. Bailey: New Encyclopedia of American Horticulture. The Macmillan Company, N.Y. 1900. 4 vols. $20.00. (By subscription only.)
3. Bailey: Talks Afield. Houghton, Mifflin & Company, Boston. 1896. $1.00.
4. Boyle: The Woodland's Orchids. Macmillan Company. 1901.
5. Campbell: The Evolution of Plants. Macmillan Company. 1899. $1.00.
6. Crozier: A Dictionary of Botanical Terms. Henry Holt & Company, N.Y. 1892.
7. Darwin: The Power of Movement in Plants. D. Appleton & Company. 1886. $2.00.
8. De Candolle: The Origin of Cultivated Plants. D. Appleton & Company. 1884. $2.00.
9. Ganong: The Teaching Botanist. Macmillan Company. 1899. $1.10.
10. Geddes: Chapters in Modern Botany. Charles Scribner's Sons, N.Y. 1893. $1.25.
11. Gray: Structural Botany. American Book Company, N.Y. 1880. $2.00.
12. Jackson: A Glossary of Botanic Terms. J. B. Lippincott Company, Philadelphia. 1900. $2.00.
13. Kerner & Oliver: Natural History of Plants. Henry Holt & Company, N.Y. 1896. $15.00.
14. Sorauer: A Popular Treatise on the Physiology of Plants. Longmans, Green & Company, London and N.Y. 1895. 9s. (about $2.50).
15. Vines: Lectures on the Physiology of Plants. Macmillian Company. 1895. $5.00.

Of the works named above, Nos. 2 and 13 are expensive and not likely to be accessible except in communities where there is a well-stocked public library. Kerner's work is written in a simple, popular style, and so profusely and beautifully illustrated as almost to explain itself without the text. No. 2, as its name implies, treats more particularly of botany in its practical relations to horticulture. No. 14 is a simple, practical treatise, easily understood, and as free from technicalities as the nature of the subject will permit.

No. 11 can be consulted with advantage. It is written in such a clear, intelligible style, and makes so plain the subjects with which it deals, that the student will find it very helpful.

HANDBOOKS

1. Britton: A Manual of the Flora of the Northern States and Canada. Henry Holt & Company, N.Y. 1898. $2.25.
2. Britton & Brown: An Illustrated Flora of the Northern States and Canada. Charles Scribner's Sons, N.Y. 1898. 3 vols. $9.00.
3. Chapman (A. W.): Flora of the Southern States. Revised ed. 1897. $4.00.
4. Coulter: A Manual of the Botany of the Rocky Mountain Region. 1885. $1.62.
5. Gray: Manual of the Botany of the Northern United States. 6th ed. Revised. 1890. $1.62.
6. Gray: Field, Forest, and Garden Botany. New ed., revised by Bailey. 1895. $1.44.
7. Willis: Practical Flora. 1894. $1.50. 3-7 by the American Book Company.
8. Watson & Brewer: Botany of California. From the United States Geological Survey. 2 vols. 1875–1880. Has been republished by the State government of California, and will be found useful to students on the Pacific slope.

Teachers desiring to do only elementary work will find the Flora in Gray's little book, "How Plants Grow," a very convenient handbook. 80 cents.

For persons not sufficiently versed in Systematic Botany to use the manuals, a number of attractive guides have been prepared, some of which are named below: —

9. Apgar: Trees of the Northern United States. American Book Company. $1.00.
10. Creevy: Flowers of Field, Hill, and Swamp. Harper & Brothers. 1897. $1.75.
11. Keeler: Our Native Forest Trees. Charles Scribner's Sons. 1900. $2.00.
12. Lounsberry (Alice): Southern Wild Flowers and Trees. Frederick A. Stokes & Company. 1901. $3.75.
13. Matthews: Familiar Trees and Their Leaves. 1896. $1.75.
14. Matthews: Familiar Flowers of Field and Garden. $1.40. Both by D. Appleton & Company.
15. Field Book of American Wild Flowers. G. P. Putnam's Sons, 1902. $1.75.
16. Newhall: The Trees of Northeast America. 1891.
17. Newhall: The Shrubs of Northeast America. 1893.
18. Newhall: The Vines of Northeast America. 1897. $1.75 each. 16–18 by G. P. Putnam's Sons, N.Y.

19. Parsons (Mrs. Dana): How to know the Wild Flowers. Charles Scribner's Sons. 1897. $2.00.
20. Wright: Flowers and Ferns in Their Native Haunts. Macmillan Company. 1901. $2.00.

PERIODICALS

The science of Botany is advancing so rapidly that a book is very soon out of date, and one who wishes to keep abreast of the current of progress should have access to one or more of the standard periodical publications dealing with the subject. Some of the most available for general use are: —

The Botanical Gazette, University of Chicago. Monthly. $4.00.

Bulletin of the Torrey Botanical Club, Lancaster, Pa. $2.00.

Forest Leaves. Pennsylvania Forestry Association, Philadelphia. Bi-monthly. $1.00.

The Plant World. Binghamton, N.Y., and Washington, D.C. Bi-monthly. $1.00.

Science. Lancaster, Pa., and Macmillan Company, N.Y. Weekly. $5.00.

Rhodora. Published by the New England Botanical Club, Boston, Mass. Monthly. $1.00.

INDEX

In the Index the numbers in Roman type (295) refer to sections: those in full-face type (**39**) refer to cuts.

Aids to Field and Laboratory Work in Botany

Apgars' Plant Analysis.   By E. A. and A. C. APGAR.

    Cloth, small 4to, 124 pages   .   .   .   .   .   55 cents

    A book of blank schedules, adapted to Gray's Botanies, for pupils' use in writing and preserving brief systematic descriptions of the plants analyzed by them in field or class work.   Space is allowed for descriptions of about one hundred and twenty-four plants with an alphabetical index.

    An analytical arrangement of botanical terms is provided, in which the words defined are illustrated by small wood cuts, which show at a glance the characteristics named in the definition.

    By using the Plant Analysis, pupils will become familiar with the meaning of botanical terms, and will learn how to apply these terms in botanical descriptions.

Apgar's Trees of the Northern United States

    Their Study, Description, and Determination.   For the use of Schools and Private Students.   By AUSTIN C. APGAR.
    Cloth, 12mo, 224 pages.   Copiously Illustrated .   .   . $1.00

    This work has been prepared as an accessory to the study of Botany, and to assist and encourage teachers in introducing into their classes instruction in Nature Study.   The trees of our forests, lawns, yards, orchards, streets, borders and parks afford a most favorable and fruitful field for the purposes of such study.   They are real objects of nature, easily accessible, and of such a character as to admit of being studied at all seasons and in all localities.   Besides, the subject is one of general and increasing interest, and one that can be taught successfully by those who have had no regular scientific training.

Copies of either of the above books will be sent, prepaid, to any address on receipt of the price by the Publishers:

American Book Company

NEW YORK   -   CINCINNATI   •   CHICAGO

Lessons in Physical Geography

By CHARLES R. DRYER, M.A., F.G.S.A.
Professor of Geography in the Indiana State Normal School

Half leather, 12mo.   Illustrated.   430 pages.   .   .   .   Price, $1.20

EASY AS WELL AS FULL AND ACCURATE

One of the chief merits of this text-book is that it is simpler than any other complete and accurate treatise on the subject now before the public. The treatment, although specially adapted for the high school course, is easily within the comprehension of pupils in the upper grade of the grammar school.

TREATMENT BY TYPE FORMS

The physical features of the earth are grouped according to their causal relations and their functions. The characteristics of each group are presented by means of a typical example which is described in unusual detail, so that the pupil has a relatively minute knowledge of the type form.

INDUCTIVE GENERALIZATIONS

Only after the detailed discussion of a type form has given the pupil a clear and vivid concept of that form are explanations and general principles introduced. Generalizations developed thus inductively rest upon an adequate foundation in the mind of the pupil, and hence cannot appear to him mere formulae of words, as is too often the case.

REALISTIC EXERCISES

Throughout the book are many realistic exercises which include both field and laboratory work. In the field, the student is taught to observe those physiographic forces which may be acting, even on a small scale, in his own immediate vicinity. Appendices (with illustrations) give full instructions as to laboratory material and appliances for observation and for teaching.

SPECIAL ATTENTION TO SUBJECTS OF HUMAN INTEREST

While due prominence is given to recent developments in the study, this does not exclude any link in the chain which connects the face of the earth with man. The chapters upon life contain a fuller and more adequate treatment of the controls exerted by geographical conditions upon plants, animals, and man than has been given in any other similar book.

MAPS AND ILLUSTRATIONS

The book is profusely illustrated by more than 350 maps, diagrams, and reproductions of photographs, but illustrations have been used only where they afford real aid in the elucidation of the text.

Copies sent, prepaid, on receipt of price.

American Book Company

New York   •   Cincinnati   •   Chicago

(112)

Lightning Source UK Ltd.
Milton Keynes UK
UKHW010638201218
334320UK00013B/805/P